Scanning Electron Microscopy of Cerebellar Cortex

Scanning Electron Microscopy of Cerebellar Cortex

Orlando J. Castejón

University of Zulia
Maracaibo, Venezuela

Kluwer Academic / Plenum Publishers
New York, Boston, Dordrecht, London, Moscow

Library of Congress Cataloging-in-Publication Data

Castejón, Orlando J.
Scanning electron microscopy of cerebellar cortex/by Orlando J. Castejón.
p. cm.
Includes bibliographical references and index.
ISBN 0-306-47711-4
1. Cerebellar cortex—Anatomy. 2. Scanning electron microscopy. I. Title.

QM455.C37 2003
612.8'25—dc21

2002043450

ISBN 0-306-47711-4

233 Spring Street, New York, New York 10013

http://www.wkap.nl/

10 9 8 7 6 5 4 3 2 1

A C.I.P. record for this book is available from the Library of Congress

Printed in the United States of America

To the memory of my father Clemente
To my mother Elba
To my wife Haydee
To my children Orlhay, Heidi, Clemente, and Julia

Il n'existe peut-être aucun centre nerveux dont la fine anatomie soit mieux connue que celle du cervelet. Et cepandant, et malgré les grand progrès réalisés en aplicant à ce theme les mèhodes de Golgi, Weigert, Erlich et les procédés neurofibrillaires, il reste encore beaucoup d'inconnues á éclaircir.

Santiago Ramón y Cajal
Trab. Lab. Inv. Biol. (Madrid), *24*, 215 (1926).

Neuroanatomists have, I think, generally recognized that the cerebellum provides the greatest challenge in our initial efforts to discern functional meaning in neuronal patterns, because there is so a beautiful geometrical arrangement of its unique neuronal constituents. Presumably, it is for this reason that we are fortunate in possessing the most refined knowledge of microstructure that is available in the central nervous system.

Sir John Eccles
The cerebellum as a neuronal machine (1967). p. 1.

Probably no other part of the central nervous system has been so thoroughly investigated and is so well known as the cerebellar cortex. For nearly a century all of its cell-types have been recognized, and the course and terminations of their processes have been described countless times by numerous authors. Yet unanimity on many doubtful points has not been reached.

Sanford Palay and Victoria Chan-Palay
Cerebellar cortex, cytology and organization (1974). p. 1.

Scope of This Monograph

This monograph represents an attempt to obtain a three-dimensional representation of the cerebellar cortex using scanning electron microscopy (SEM) and related techniques. In the first stage, we tried to obtain a three-dimensional picture of the cytoarchitectonic arrangement of the cerebellar cortex. A second effort then was directed to characterizing the intracortical circuits formed by mossy and climbing fibers and the intrinsic cerebellar circuits formed by cerebellar neurons.

Chapter 1 reviews the materials and methods used. Chapter 2 deals with the cerebellar white matter and characterization of afferent mossy and climbing fibers. Chapter 3 describes granule cell organization. Chapters 4, 5, 6, and 7 examine mossy fibers, Golgi, Lugaro, and unipolar brush cells. Chapter 8 is devoted to Purkinje cells. Chapter 9 describes climbing fibers and their synaptic connections to the granule cell, Purkinje cell, and molecular layer. Chapters 10 and 11 deal with basket and stellate cells. Chapter 12 reviews the three-dimensional morphology of cerebellar neuroglial cells, mainly oligodendrocytes and Bergmann cells. Chapter 13 deals with cerebellar capillaries. And finally, Chapter 14 summarizes the contribution of SEM to cerebellar neurobiology.

A complete morphological three-dimensional approach for studying the cerebellar cortex must rely on the use of multiple and complementary techniques. For this purpose, the reader is referred to our original articles on correlative microscopy of cerebellar cortex. The scope of the present monograph deals only with the application of SEM and related sample preparation techniques to the study of vertebrate cerebellar cortex.

Acknowledgments

I express my sincere thanks to the following Venezuelan and International Scientific Institutions that made it possible to conduct research on scanning electron microscopy of the vertebrate cerebellar cortex, during the period 1976–2002: Instituto de Investigaciones Biológicas "Dres. Orlando J. Castejón y Haydee Viloria de Castejón," Facultad de Medicina, Universidad del Zulia, Maracaibo, Venezuela; Yerkes Research Center and Department of Chemistry, Emory University, Atlanta, Georgia, U.S.A.; and the Canadian Department of Agriculture, Ottawa, Canada. My deep gratitude to the warm and constant encouragement of Prof. Geoffrey Haggis and Dr. Robert Apkarian from these institutions. My special thanks to the collaborative research work of my wife Haydée V. Castejón. I am very much indebted to Nelly Montiel for her enthusiastic participation with scanning electron microscope sample preparation techniques and to José Espinoza for skillful maintenance and operation of our electron microscopes. My special thanks to Dr. Consuelo Valero for the preparation of human cerebellar cortex. To Dr. Robert Apkarian for his continued help with the field emission scanning electron microscopy (FESEM) at Emory University. Photographic assistance of Ralph Caspersen is also acknowledged. It is also a pleasure to express my deep gratitude to Laura Villamizar for her competent secretarial assistance. Grateful thanks are expressed to Drs. Rogério Monteiro (Institute of Biomedical Sciences, Porto University, Portugal), José Mascorro (Department of Structural and Cellular Biology, Tulane University School of Medicine, New Orleans), Kenneth Moore (Electron Microscopy Center, Iowa University, Iowa City), and Haydée V. Castejón (Biological Research Institute, Faculty of Medicine, University of Zulia) for their meticulous review and linguistic improvement of the manuscript.

I also express my sincere thanks to the following journals for their generosity in giving permission for reproductions of figures: *Cell and Tissue Research*, *Scanning*, *Biocell*, and *Journal of Submicroscopic Cytology and Pathology*. It has been a pleasure for me to work with the staff of Kluwer Academic Publishers, specially with Kathleen Lyons (Senior Editor, Bioscience) for the editing process of this monograph. It is a great pleasure to have such excellent cooperation and understanding.

Preface

For all early neuroanatomists devoted to the microstructure of cerebellar cortex self-made drawings were the unique documentation to illustrate their discoveries and findings. Ramón y Cajal (1911, 1955) published the most impressive collections of drawings in his monumental treatise "Histologie du Système Nerveux de L'Homme et des Vertébrés," translated to the French language by Dr. L. Azoulay and edited by the Consejo Superior de Investigaciones Científicas (Madrid). Spectacular Golgi light microscopic (LM) observations were published by Ramón y Cajal (1888–1926); Golgi (1874–1886); Denissenko (1877); Van Gehuchten (1891); Lugaro (1894); Retzius (1894); Terrazas (1897); Bielschowsky and Wolff (1904); Estable (1923); Jakob (1928); Pensa (1931); Fox et al. (1954–1959); and Scheibel and Scheibel (1954).

The advent of electron microscopy initiated a new era of correlated light and electron microscopic observations on the cerebellar cortex. Gray (1961), Palay (1956–1974), Fox et al. (1964–1967), Hámori and Szentágothai (1964–1968), Mugnaini et al. (1974–1994), Hillman (1969), Sotelo (1969), and Uchizono (1965–1969), among others, published new bidimensional illustrations and drawings on the structure of cerebellar cortex.

After such numerous and elegant investigations, some members of the neuroscience community asked: What's new on cerebellar microstructure? What is the real contribution of the scanning electron microscope (SEM)? This monograph is an attempt to answer these questions. The microstructure of the cerebellar cortex is not an extinguished subject. Modern three-dimensional microscopic techniques, such as SEM, confocal laser scanning microscopy (CLSM), and related immunocytochemistry methods and computer-assisted microscopy with different levels of resolution and magnification provide new findings on the cerebellar structure and organization.

In the present volume we describe the contribution of SEM to the study of vertebrate cerebellar cortex including humans. This monograph represents a systematic effort over more than three decades directed to demonstrate the potential contribution of SEM in the study of the three-dimensional organization of the central nervous system.

This book is directed to young neuroscientists fascinated with the study of the cerebellar cortex. I hope the new generation will understand my intent in writing it: to provide a refined three-dimensional representation of the cerebellar cortex.

Orlando J. Castejón

Contents

Chapter 1

Sample Preparation Methods for Scanning Electron Microscopy

INTRODUCTION

In order to properly interpret images of nerve cells observed in the scanning electron microscope (SEM), it is of fundamental importance that fixation, dehydration, critical point drying (CPD), and metal deposition be accompanied with minimal distortion of nerve and glial cell structures and their cytoarchitectonic arrangement within a gray center. It is also of importance that any artifacts which are ever produced be recognized.

Nerve tissue is very soft with a high water content; therefore, it is subjected to significant dimensional and conformational changes during the preparative procedures for SEM (Boyde, Bailey, Jones, & Tamarind, 1977). However, it should be considered that these changes could, in certain regions where the narrow extracellular space has been enlarged, expose hidden neuronal surfaces and reveal interneuronal relationships.

The purpose of this chapter is to provide an introductory understanding of the basic principles, theory, usefulness, limitations, and artifacts of the preparative procedures used for SEM examination of nerve tissue. The instructions for the preparative techniques are directed to those students and young investigators without previous experience to enable them to prepare specimens without technical supervision. In addition, I have critically reviewed each basic preparatory step prior to SEM examination. I have included some special procedures that I personally have found to be useful. This chapter is organized in several introductory sections dealing with fixation, dehydration, critical point drying (CPD), mounting and orientation of nerve specimens, and metal deposition. The results and effects of various preparative procedures for nerve tissue are described (in a teaching manner) and analyzed emphasizing advantages, limitations, and sources of artifacts.

CONVENTIONAL SEM TECHNIQUE OR SLICING TECHNIQUE

Conventional SEM technique or slicing technique includes the following preparatory steps: fixation, trimming and obtaining of nerve tissue slices, dehydration, CPD, specimen mounting with orientation, and metallic coating.

The cerebellum was chosen as a model organ from the the central nervous system and nerve tissue preparation studies for the following reasons:

1. Its geometric arrangement in a three-layered structure can be identified easily by light microscopy (LM), transmission electron microscopy (TEM), and SEM.
2. The cerebellum has the simplest cytoarchitectonic arrangement of the central nervous system, especially when compared with the cerebral cortex.
3. The cerebellum is a well-known structure from neuroanatomical, neurochemical, and neurophysiological points of view.
4. The presence of two macroneurons (Purkinje and Golgi cells) and five microneurons (granule, unipolar brush, Lugaro, basket, and stellate cells) are distinguished easily in different neighboring layers, thus offering immense possibilities for studying neuronal geometry, synaptic connections, and intracortical circuits.
5. Different types of synaptic junctions are present: large mossy multisynaptic complexes (mossy glomerular regions), axosomatic synapses (especially upon Golgi, Purkinje, and stellate cells), and spine synapses in the molecular layer (mainly climbing and parallel fiber–Purkinje spine synapses).
6. The cerebellum has a constant synaptic organization throughout the whole vertebrate phylogenetic scale.

FIXATION PROCEDURES

Primary Fixation with Glutaraldehyde by Immersion Technique

Optimal fixation of a gray center is the basic criterion for preserving nerve cell volume and, therefore, three-dimensional configuration and spatial relationships in situ (Figure 1). Glutaraldehyde, commonly used as the primary fixative, generally causes shrinkage of nerve cells, mainly because slightly hypertonic fixatives (320–350 mOsm) are used. This nerve cell volume alteration is expressed as a 10–20-nm widening of the membrane-to-membrane extracellular space between neighboring nerve cells.

The total osmolarity of the fixative given by the osmotic pressure of the glutaraldehyde fixative and the buffer solution (Lee, 1983) is an important factor to be considered in nerve cell volume alteration. In addition, the type of buffer used, either phosphate or cacodylate, also is important. In our experience, 3–5% glutaraldehyde–0.1M phosphate buffer solution induced 10–20% more shrinkage of nerve cells than a similar fixative concentration in cacodylate buffer, apparently due to extractive properties of the phosphate buffer on the cell matrix.

The immersion fixation technique has been used in our laboratory for human pathological material. For ethical and practical reasons, this seems the most appropriate technique for sample stabilization and preservation. However, the immersion fixation procedure is not optimum for preserving the neuron–glial cell relationship in situ.

Postfixation in Osmium Tetroxide

Osmium tetroxide (OsO_4) should be considered as a secondary fixative to stabilize tissue and membrane lipids since glutaraldehyde fixation only cross-links proteins. Most

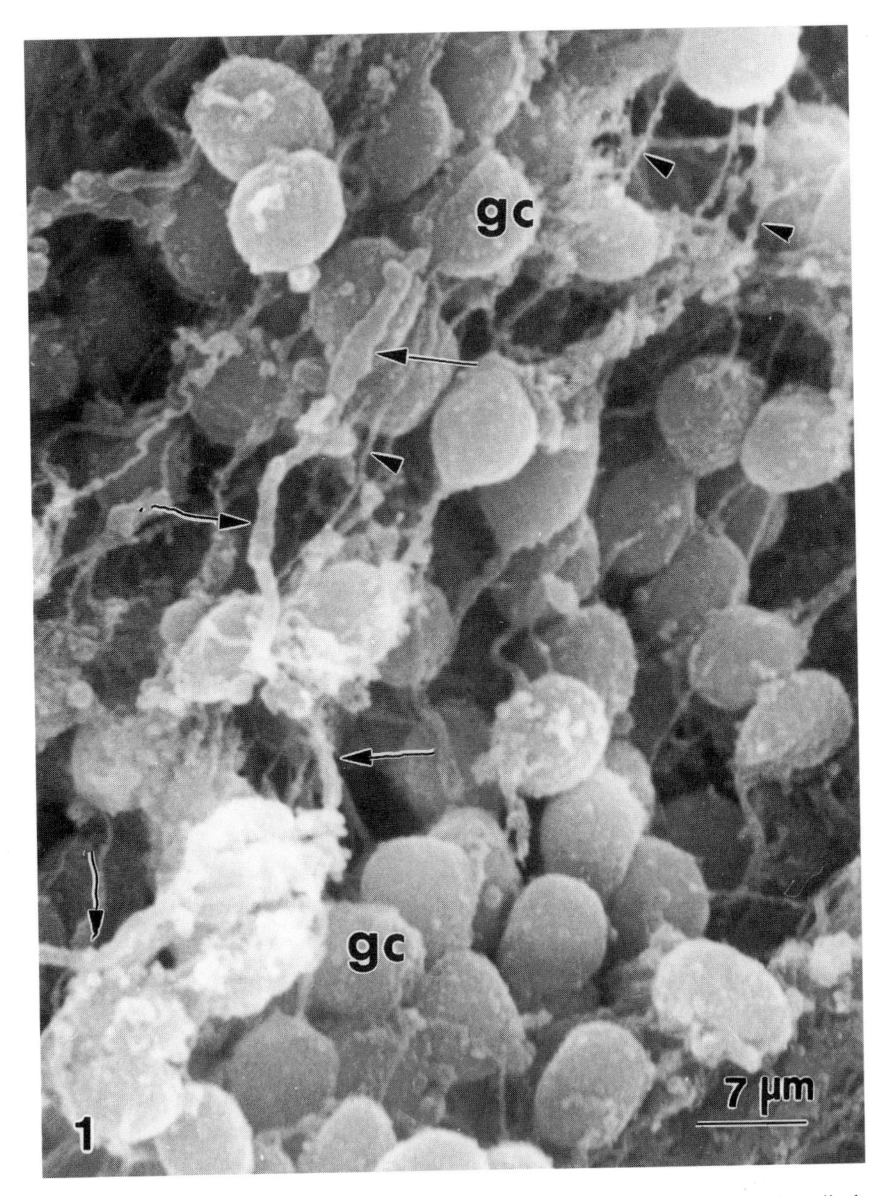

FIGURE 1. Scanning electron microscopy slicing technique. Teleost cerebellar granule cells fixed by vascular perfusion with 2% glutaraldehyde–0.1M phosphate buffer solution, pH 7.2, osmolarity 320 mOs/L. Postfixation in similarly buffered 1% osmium tetroxide (OsO_4). Good preservation of granule cell (gc) surface morphology, cell volume, and processes is observed. Afferent mossy fibers (arrows) and their collaterals (arrowheads) can be accurately traced through the width of the granular layer.

investigators choose to postfix cells with OsO_4 because it stabilizes lipid containing structures against extraction by organic solvents. Because OsO_4 also acts as an emitter of both secondary and backscattered electrons in SEM, it contributes either to stabilization or visualization of cells (Gamliel, 1985). Thus, in the preparation of SEM samples, OsO_4 not only preserves lipids, but also increases the mass density of the tissue and the generation of type II and III secondary electrons (SEs). Postfixation in OsO_4 is necessary to stabilize the plasma membrane against the inevitable tensions that occur during dehydration and CPD (Arro, Collins, & Brunk, 1981). Whether OsO_4 fixation, when used as a secondary fixative, causes additional shrinkage is difficult to ascertain. In conventional electron microscopy (EM), OsO_4 is commonly used in 0.01M veronal, collidine, or cacodylate buffers. It might be expected that electrostatic interactions of charged protein particles and osmotic forces could induce swelling or shrinkage during OsO_4 fixation (Tooze, 1964).

When glutaraldehyde was used as the primary fixative, postfixation in OsO_4 was shown to cause swelling in rat brain tissue (Eins & Wilhems, 1976). Brunk, Bell, Colling, Forsby, and Fredriksson (1975) studied the effects of glutaraldehyde, at different effective osmotic pressures, on human glial cells in vitro with time-lapse cinematography and SEM. These authors reported that optimal preservation of fine structure with glutaraldehyde fixative was obtained when the osmolarity of the vehicle was approximately 300 mOsm/L. Variation in the concentration of glutaraldehyde did not cause any visible changes in cell morphology, although it did affect the stability of swelling artifacts. According to Boyde (1978), neurons fixed with 1% OsO_4 or 5% glutaraldehyde in 0.0475M sodium cacodylate did not shrink.

Since glutaraldehyde fixation could initially be a reversible process and the cell membrane remains semipermeable to most solutes after glutaraldehyde fixation, it is recommended that the osmolarity of OsO_4 fixative be maintained the same as that of the wash buffer after glutaraldehyde fixation in order to reduce volume alterations (Lee et al., 1982).

Glutaraldehyde-Fixation Artifacts

Blebs or blister formation has been observed in glutaraldehyde fixation for TEM and SEM studies (Castejón, 1993; Shelton & Mowczko, 1978).

THE PREFIXATION STATE OF THE NERVE TISSUE

Prior to immersion fixation, small animals such as mice and rats should be anesthetized and decapitated and the brain tissue exposed through the skull. This treatment undoubtedly introduces chemical and mechanical alterations in the fine structure of neurons, glial cells, synaptic contacts, and blood–brain barrier. After delicate dissection and sectioning of cranial nerves and ultimate sectioning of the brain stem, the whole brain is extracted from the skull. This procedure takes from 5 to 20 min depending on the vertebrate species and the individual's skill. In all cases, this procedure causes trauma, anoxia, and brain ischemia, which induce brain edema (Castejón, 1980). In some cases, the samples should be washed in buffer solution to eliminate the extravasated blood, which could deteriorate the primary fixative. After isolation, the samples must be cut into small tissue blocks, 2–3 mm thick, using a sharp razor blade. The trimming procedure induces additional deformation and stress forces. Since the sample is still unfixed, the anoxic period is prolonged and autolytic changes are initiated. The tissue block should be gently handled with forceps or suspended with a wooden stick and immersed in freshly prepared 3–5%

glutaraldehyde in 0.1M phosphate buffer solution for 2 hr. Oxygenation of the solution is also recommended. As the fixative penetrates the tissue, the concentration of the fixing compound is reduced due to a loss by reaction with tissue components (Lee, 1983). Therefore, the concentration of the fixative compound should be high. Since cell matrix extraction may occur, overfixation is not recommended. A brownish coloration and an increase in the consistency of the tissue pieces are observed after two hours of fixation. Several rinses in the buffer solution are strongly recommended before postfixation in OsO_4.

VASCULAR PERFUSION FIXATION TECHNIQUE

Due to the high sensitivity of nerve tissue to trauma and anoxia, this technique has more advantages than the immersion technique for in situ preservation of cytoarchitectural arrangement of the gray centers. A large table is used for rabbits, guinea pigs, or monkeys. The perfusion equipment consists mainly of a perfusion flask (1 L capacity), perfusion (surgical rubber) tubing (6 mm inside diameter), and Pasteur glass pipettes heated and drawn into thin glass puncture micropipettes similar to the glass microelectrodes used in neurophysiology for intracellular recordings (Figures 2 and 3). They are used as needles to puncture the left cardiac ventricle (Figure 4). These modified glass micropipettes are connected to the perfusion flask by means of the perfusion tubing. A manual valve controls the flow of the fixative. The height of the fixation bottle, about 200 cm should be adapted according to the physiological pressure in the brain vessels of the experimental animal. To avoid higher pressures in the vascular system during the perfusion process, an opening is made in the right auricle. A respiratory apparatus consisting of a tank of 95% oxygen (O_2) and 5% carbon dioxide (CO_2) is used. A rubber tube provided with a micropipette connects the tank to the hole opened in the trachea of the animal. To observe the arrival of the fixative to the central nervous system, an observation window should be opened in the skull. A constant volume of the fixative should descend from the perfusion flask and stream into the vascular bed by gravity. Clogging of blood cells by the fixative solution should be prevented by previously flushing the vasculature with a phosphate buffered solution. The duration of the perfusion is determined by the flow of the fixative free of blood through the right auricle, as well as by observing the yellowish coloration appearing in the nerve tissue, and in other organs such as liver, skin, and muscles. Before and during the vascular fixation, the animal receives a mixture of 95% O_2 and 5% CO_2 by means of a glass pipette inserted into the trachea. After completing the vascular perfusion, the skull bones are gently cut using a dental drill (Figure 5). After very careful dissection of cranial nerves, the brain stem is sectioned at the level of the cervical spinal cord. Figure 6 shows the isolated and fixed central nervous system. After vascular perfusion and removal, the brain should be placed in fresh fixative solution by immersion "in toto," cut into 2–3 mm segments, and immersed in fresh fixative. The reader is referred to Palay and Chan-Palay (1974) for detailed description of the procedure of the vascular perfusion technique. After primary vascular fixation with glutaraldehyde, postfixation by immersion in OsO_4 should be done. The quality of fixation by vascular perfusion should be initially evaluated with the LM. The capillaries should be free of blood cells (Figure 7) and the surrounding pericapillary tissue should be well preserved.

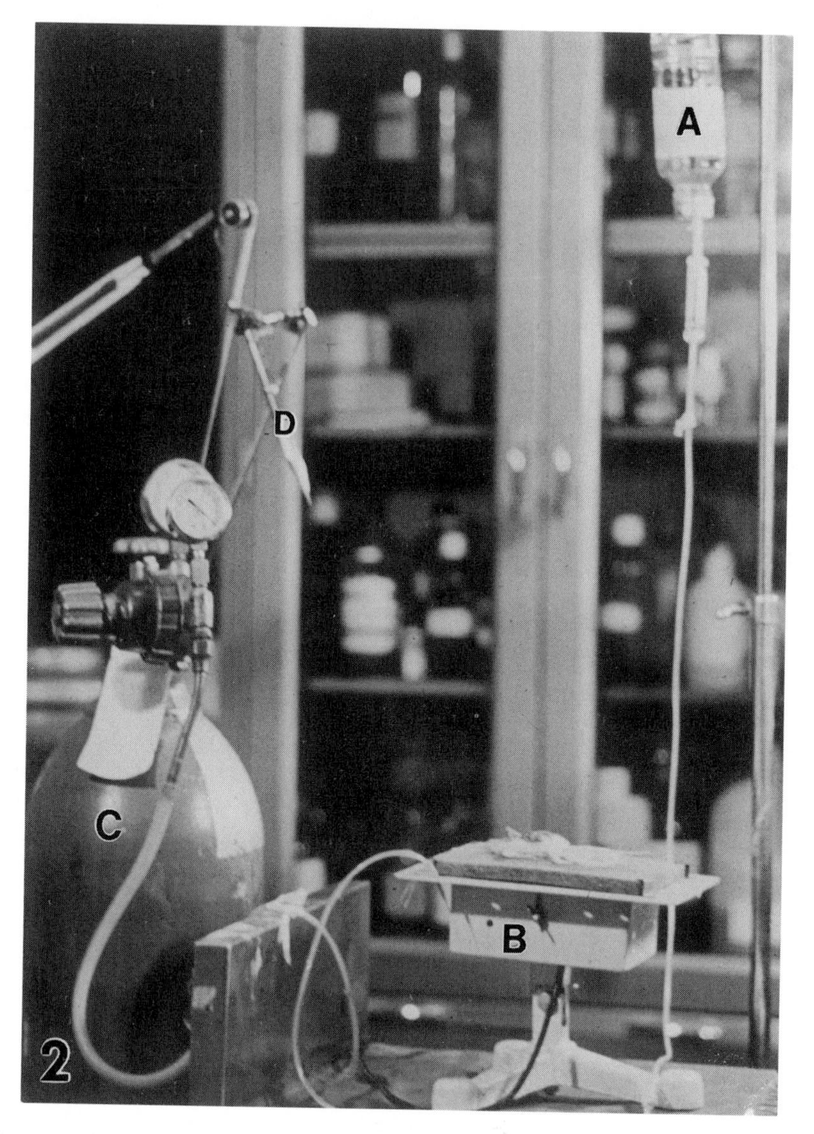

FIGURE 2. Vascular perfusion equipment. A: perfusion flask. B: surgical table for rodents. C: cylinder containing a mixture of 95% O_2 and 5% CO_2. D: driller for opening the skull.

CRITERIA FOR GOOD FIXATION AND OPTIMAL PRESERVATION OF NERVE TISSUE

The criteria for optimal preservation and fixation of nerve tissue are as follows:

1. smooth outer surface of nerve cells as shown in Figure 1 and conservation of neuronal geometry;

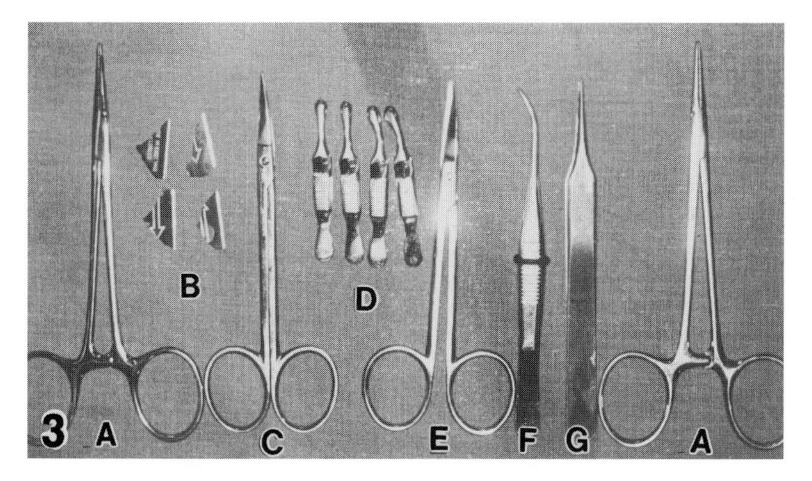

FIGURE 3. Surgical instruments used for isolating the mouse brain. A: Kogel forceps. B: razor blade fragments. C: curved scissor. D: small surgical clip. E: right scissor. F: curve dissecting clip. G: fine point tweezer.

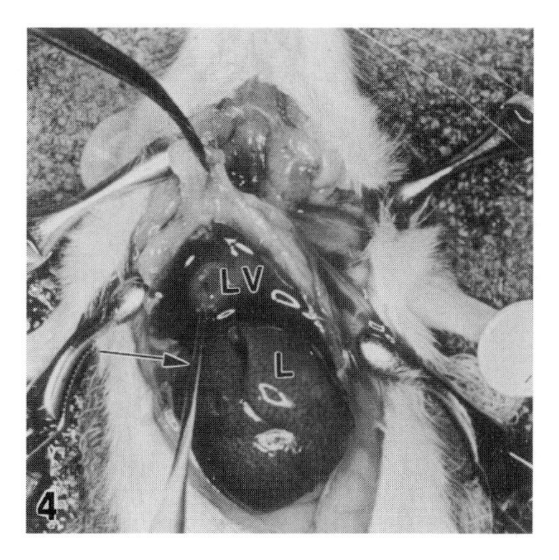

FIGURE 4. Albino mice perfused with glutaraldehyde. The glass micropipette (arrow) is inserted into the left ventricle (LV). The perfusion process can be followed by observing the change of coloration in the liver (L).

2. absence of widened extracellular spaces between adjacent nerve cells indicating that significant volume and dimensional changes (no more than 10–15%) have not occurred;
3. continuity of nerve cell processes with the nerve cell soma in unfractured surface areas;
4. intact topographic relationship of nerve cell dendritic processes with afferent axonal fibers, at the level of synaptic junctions; and
5. maintenance of the neuron–glial cell unit topographic relationship in macroneurons provided with satellite neuroglial cells.

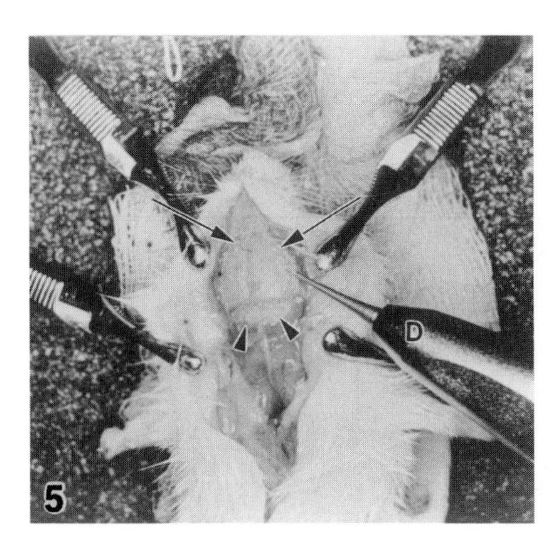

FIGURE 5. Opening of mouse skull using a fine drill (D). The cerebral hemispheres (arrows) and the cerebellum (arrowheads) are distinguished.

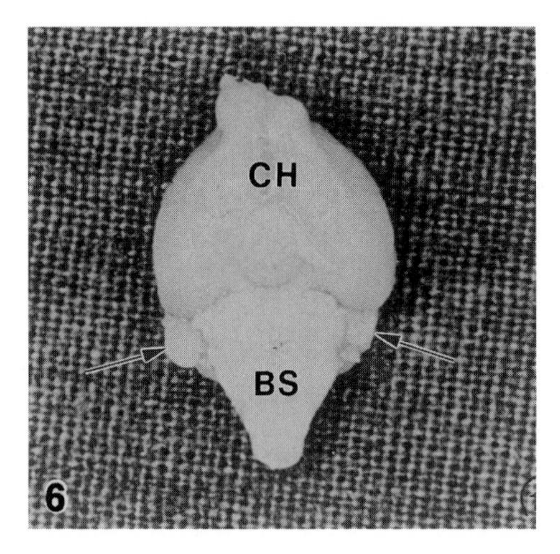

FIGURE 6. Isolated fixed mouse central nervous system showing the inferior surface of the cerebral hemispheres (CH), cerebellum (arrows), and brain stem (BS).

TRIMMING PROCEDURE AND OBTAINING NERVE TISSUE SLICES

After fixation, the nerve tissue fragments should be trimmed to slices no larger in diameter than the support stub and as thin as possible (1–2 mm). During the trimming process, the nerve tissue is oriented on the support stub to assure that sagittal and transverse sections of the cortical structures of the gray centers are obtained. During the trimming procedure air drying should be avoided by keeping the nerve samples immersed in the buffer solution. Trauma and unnecessary manipulation of the tissue fragments should also be avoided.

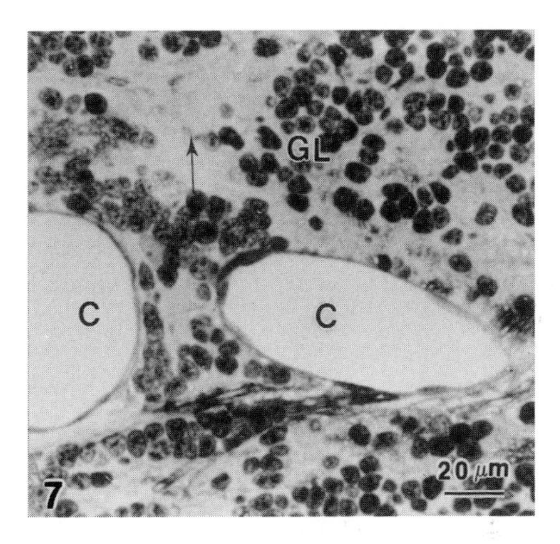

FIGURE 7. Light microscope image of perfused granular cerebellar layer (GL) showing capillaries (C) with lumens devoid of blood cells.

DEHYDRATION

Standard dehydration schedules inevitably cause cells and tissues to swell (at lower solvent concentrations) and then to shrink (Boyde et al., 1977), often with the production of cracks. Thus, dehydration schedules should omit the lower solvent concentrations and begin with 70% ethanol to avoid severe volume shrinkage. Shrinkage may also alter the three-dimensional relationship of cells to each other (Bell, 1984). This problem is of paramount importance in the study of nerve tissue organization especially the neuron–glia unit and the relationship between afferent fibers and neuronal cell layers. In nerve tissue preparation, and because further treatment of nerve tissue during cryofracture and thawing requires the use of absolute ethanol, we routinely use ethanol as a dehydrating solvent. Ethanol produces a better preservation of cytoplasmic matrix, endoplasmic reticulum, mitochondria, and nuclei when compared with other dehydrating agents (Lee, 1983).

CRITICAL POINT DRYING METHOD

The CPD method (Anderson, 1951) is the commonly used technique for removing the liquid phase from cerebellar nerve tissue. This method avoids the damaging surface tension forces associated with solvent evaporation drying, by taking a liquid (CO_2) to its critical point (Boyde & Maconnachie, 1981; Cohen, 1977), where surface tension is minimal.

Due to its high water content, nerve tissue might collapse or become damaged during CPD. Water plays a structural role in the macromolecular architecture of neurons and neuroglial cells. These cells maintain their shape by the water molecules that are intercalated amongst the macromolecular network of plasma membrane, cytomembranes, and cytoskeleton. As the water molecules are withdrawn during dehydration or CPD, the nerve cells experience volume and surface changes. Therefore, delicate and gentle specimen preparation procedures are used to reduce the volume and surface stresses to 10–15%. The CPD method essentially avoids surface tension effects (Cohen, 1974). Theoretical and general practical considerations concerning the CPD method have been discussed by Boyde and Wood (1969), Boyde (1972), Nemanic (1972), Lewis and Nemanic (1973), and Cohen (1974, 1979).

In our laboratory, absolute ethanol and liquid carbon dioxide (CO_2) are used as intermediate and transitional fluids, respectively, in the CPD process. Liquid CO_2 has a critical temperature of 31.1°C and a critical pressure of 72.9 atm (55,400 Torr) (Anderson, 1951). Specimen containers for transferring the sample to the CPD apparatus should be selected according to the dimensions of the nerve sample.

Before CPD, the nerve tissue should be placed in a closed wide-mouth jar and immersed in absolute ethanol. The nerve specimen is placed in the CPD apparatus (Figure 8), below the critical temperature. When the system is carried through the critical point to a temperature above the critical temperature, the nerve tissue is in a gaseous phase without ever passing through the gas–liquid interface. By keeping the temperature above the critical point, the valve can be opened to air and the pressure allowed to drop slowly to atmospheric pressure. The specimen is now critical point dried. During the CPD process, the intermediate or dehydration fluid (absolute ethanol) is completely miscible with the transition fluid (liquid CO_2).

The main artifact caused by the CPD method seems to be shrinkage—a 20% reduction in linear dimensions after this procedure is acceptable (Gamliel, 1985). Wrinkling of the cell membrane may be due to shrinkage of internal structural components. Small holes in the cell membrane, which sometimes look like circular ruptures, can also be related to swelling and shrinking events that occurred during the conventional CPD process (Figure 9).

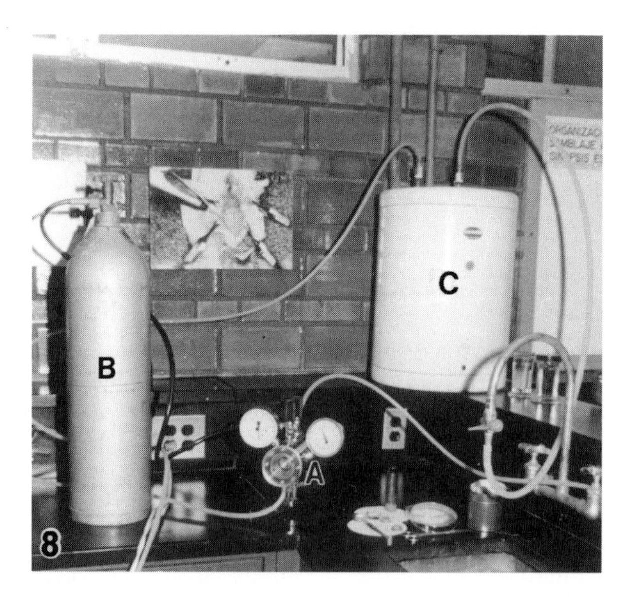

FIGURE 8. Critical point drying equipment. A: critical point dry apparatus. B: liquid CO_2 cylinder. C: hot water apparatus.

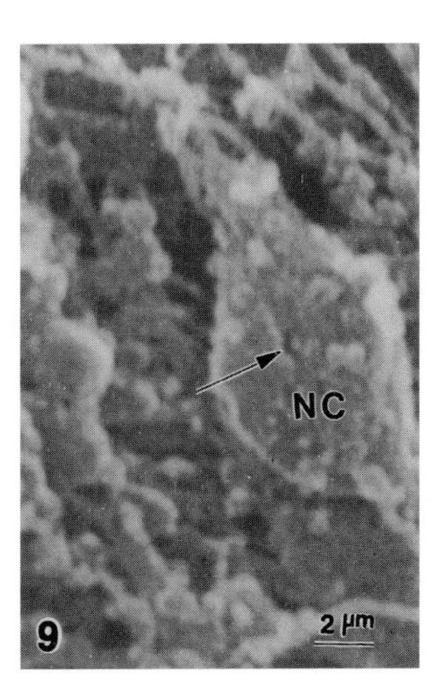

FIGURE 9. Artifactual damage (arrow) induced in a nerve cell (NC) by the shrinkage during conventional critical point drying.

SPECIMEN MOUNTING AND ORIENTATION

Nerve tissue fragments can be mounted and oriented on specimen stubs with double-sticky pressure sensitive adhesive tape or directly attached with conductive paint or paste. A binocular dissecting microscope should be used to observe the cut or fractured surface. The unuseable part of the specimen can be set into the cement. The tissue fragment should be gently oriented according to the sagittal, transverse, or tangential position desired. A moment should be allowed until the adhesive sets.

METAL DEPOSITION

The most important reason for coating the nerve tissue is to increase its thermal and electrical conductivity. The thin layer of metal, which is usually applied to insulators to make them conductive, is also the source of the majority of SEs. The rationale and mode of application of thin films to nonconducting material has been elegantly described by Echlin and Hyde (1972) and Echlin and Kayes (1979). Figure 10 illustrates the sputter coater used in our laboratory for gold–palladium coating.

To overcome the limitations on resolution in bulk samples by the range of the electrons collected in the signal, a metal coating must be thinner than the range of the critical signal electrons (Everhart & Chung, 1972). This has been accomplished in nerve tissue by applying a 5–10 nm gold–palladium coating, for conventional SEM, or 1–2 nm chromium coating, for either conventional or high resolution SEM (Castejón, 1991). Since the images dealt with are largely derived from the metal film and not from the actual nerve cell surfaces, relevant properties of the metallic coatings affecting the image should be considered.

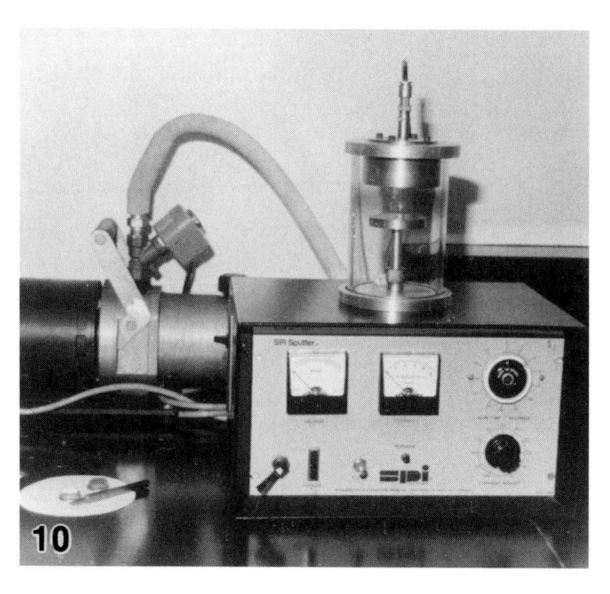

FIGURE 10. Sputter coater used for gold–palladium coating.

There are three main categories of electrons which leave the surface of a solid when it is struck by a beam of primary electrons of sufficient kinetic energy: (1) the elastically reflected or backscattered electrons which form a sharp energy distribution peak close to the primary beam energy; (2) inelastically scattered electrons which have lost discrete amounts of energy before escaping from the surface; and (3) the "true" SEs which form the bulk of the emitted electrons (Echlin & Hyde, 1972). These latter electrons originate from within 5–10 nm of the surface and are used in the secondary or emissive mode of operation in conventional SEM. With a primary beam of about 10 kV, a 10-nm thick layer of a heavy metal alloy, such as gold–palladium, would significantly improve the electron emission from the nerve tissue. The incident electrons penetrate the sample for some distance, which are a function of both accelerating voltage and the atomic number of the specimen being examined. This penetration gives rise to a variety of signals. At the point of incidence and in the first scattering event, type I SE signals are produced. Backscattered electrons then emerge from the specimen at some distance from the point of entry and produce type II SE signals at the specimen surface, or they may leave the specimen and produce type III SE signals within the microscope chamber (Peters, 1985b). Type II and III SE signals dominate in conventional SEM and type I SE signal dominates the image in the modern, analytical high resolution SEM. Surface information depends on surface topography, the energy of the primary electrons, and the mass density of the specimen. The reader is referred to Peters (1985b) for a detailed description of signal collection strategies. Figure 11 shows the cut and fractured surfaces of a cerebellar specimen. The triple layered structure of cerebellar cortex can be vaguely distinguished in the SEM, at low magnification.

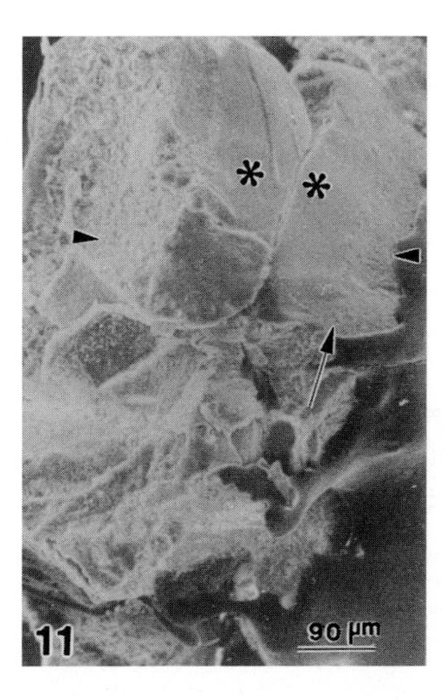

FIGURE 11. Fragment of cerebellar cortex faintly showing a triple layered structure: granular layer (arrowhead), Purkinje cell layer (arrow), and molecular layer (asterisks).

For conventional SEM, we use a JEOL 100B TEM with an ASID scanning attachment equipped with a tungsten filament source operated at 10, 20, or 60 kV, thus ensuring a high SE image with a resolution better than 10 nm. At these instrument parameters, type II SEs are predominantly collected from the gold–palladium coated nerve cells, with a nominal topographic resolution of about 30–50 nm. But depending on practical interference problems, such as sample contamination, mechanical vibrations, or stray fields (Black, 1974) the resolution could deteriorate to about 100 nm. For conventional SEM, low and intermediate magnifications are used. High accelerating voltage is applied to establish the smallest beam diameter, highest gun brightness, and to reduce the background signal contrast.

NERVE CELL SPECIMENS COATED WITH THICK GOLD–PALLADIUM FILMS

Nerve tissue conventionally prepared for SEM is coated with a 5–10 nm thick gold–palladium film to increase electron conductivity and surface contrast (Echlin & Hyde, 1972). Metal coating with 5 nm of gold–palladium alloy produces sufficient secondary electrons to avoid charging effects and yields a satisfactory signal/noise ratio resulting in a resolution in the 100–200 nm range. A metal thickness of 20 nm produces a smudged appearance and an obvious loss of resolution. The thickness of the metallic coating limits the resolution and obscures cell surface details. The morphology of the exposed

FIGURE 12. Turbo-pumped sputter deposition system for chromium coating (courtesy of R.P. Apkarian).

or hidden nerve cell surfaces are revealed, but the visualization of subsurface details of the glycocalyx layer is limited. In order to improve resolution, thinner metal films, such as of chromium, should be used, which potentially allows the visualization of the glycocalyx layer (Castejón, 1991) and the true nerve cell outer surface. Figure 12 illustrates a turbo pumped sputter deposition system used for chromium coating.

Gold–palladium produces good contrast at low and medium magnifications. The metal film is smooth at the outer surface of the nerve cell soma, dendrites, and axons when the slicing technique is used (Figure 1). With the freeze–fracture method, which demonstrates the inner surface cytoplasmic and nuclear details, the metallic film evenly coats cytomembranes, cell organelles, cytoskeleton, and nuclear chromatin. The method also shows distinct topographic substructures, particle aggregates, and the spatial relationship between nuclear and cytoplasmic compartments, which are sharply outlined and in close apposition.

Nerve Cell Surfaces Decorated with Gold–Palladium

Very thick gold–palladium accumulation distorts the topography and also produces the redistribution of metal during its deposition, thereby generating undesirable decoration of surface features (Figure 13). The nerve cell surface presents a distinct microroughness contrast and a granular appearance. Therefore, the metallic layer should be as thin as possible in order to coat rather than to decorate.

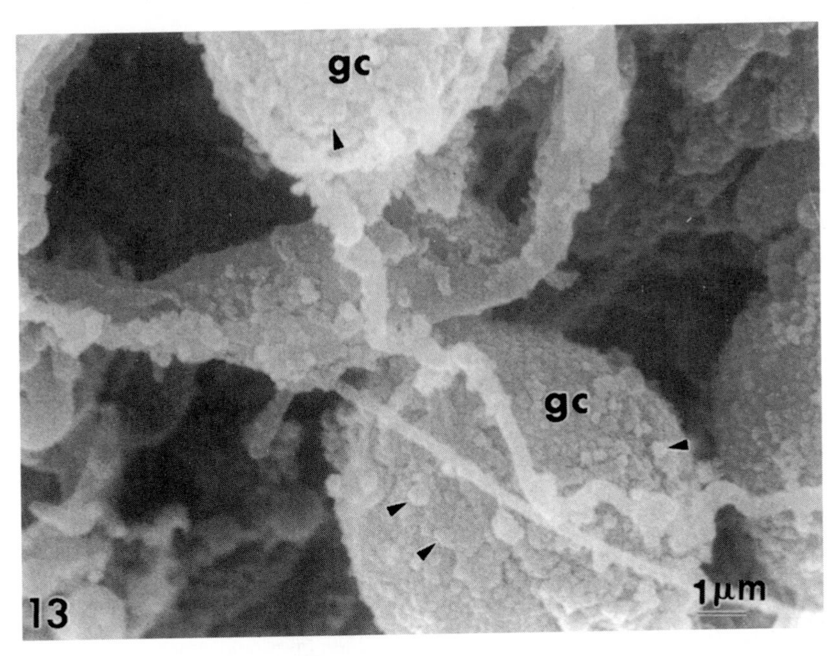

FIGURE 13. Rough granular outer surface of granule cells (gc), coated with a heavy deposit of 10–20 nm thick gold–palladium as a decorative artifact (arrowheads) that masks the true outer neuronal surface.

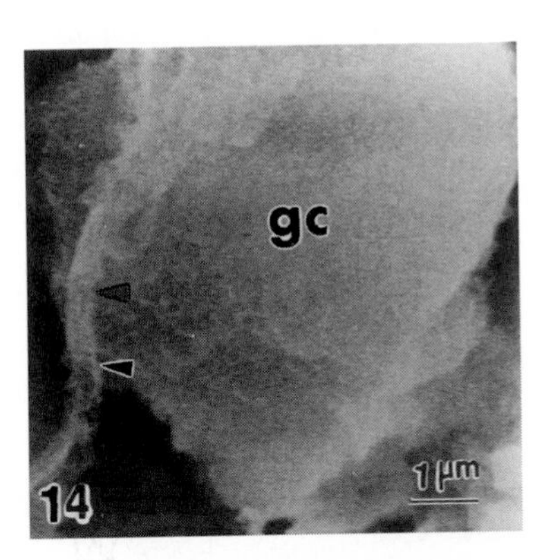

FIGURE 14. Field emission scanning electron microscopy (FESEM) micrograph of a granule cell (gc) coated with a thin layer of chromium, 1–2 nm thick. A smooth true outer surface corresponds to the glycocalyx layer. The axon initial segment is also seen (arrowhead).

Nerve Cells Specimens Coated with Thin Chromium Films

Thin chromium films (1–2 nm thick) produce a different outer surface morphology of nerve cells, characterized by a smooth, cotton-like surface corresponding to the glycocalyx layer (Figure 14). The material contrast is lower but the resolution is notably improved, especially at low and intermediate magnifications. The majority of the contrast is generated by type I and II SEs (SE-I and SE-II type signals) (Apkarian & Joy, 1988; Peters, 1985b).

SPECIAL SEM PREPARATION TECHNIQUES

Ethanol-Cryofracturing Technique

We have widely employed the ethanol-cryofracturing technique originally designed by Humphreys et al. (1974, 1975), for liver and kidney tissue, to image the hidden outer surfaces of human cerebellar neurons and neuroglial cells (Castejón, 1988, 1990a; Castejón & Caraballo, 1980a, 1980b; Castejón & Castejón, 1987). Samples of 3–5 mm thick human nerve tissue are fixed by immersion for 2–16 hr in 4% glutaraldehyde-phosphate buffer solution, 0.1M, pH 7.4. The fixed tissue is cut into pieces about 1 × 2 mm and dehydrated in ethanol. In the final change of absolute ethanol, the pieces are inserted into small cylinders of parafilm filled with absolute ethanol. Air bubbles should be avoided within the cylinder. Both ends of the cylinder should be clamped shut with a forceps. The cylinder is seized held with a forceps in liquid nitrogen until frozen; then stored in a Petri dish with liquid nitrogen prior to cryofracture. The fracture is made with a Smith–Farquhar tissue sectioner (Figure 15) fitted with a precooled razor blade. The fragments are later placed in ethanol for thawing at room temperature. The samples are processed according to the technique of Humphreys, Spurlock, and Johnston (1974, 1975) with minor modifications (using phosphate buffer instead of cacodylate buffer). CPD is done with liquid CO_2, followed by coating with 10–20 nm thick gold–palladium. The cryofracture process exposes the hidden outer and inner surfaces of neurons and neuroglial cells (Figure 16) and the afferent and intrinsic fibers (Figure 17) allowing visualization of the cytoarchitectonic arrangement of the cerebellar cortex. The fracture line always follows the plane of tissue weakness, apparently represented by the satellite neuroglial cells. The neuroglial cells ensheathing the nerve cells are easily removed thus exposing the hidden outer somatic surface of nerve cells. At the sites previously occupied by satellite neuroglial cells large crevices are observed. Freezing with liquid nitrogen (medium freezing rate) produces a cryodissection of nerve cell processes, which are kept intact due to the freezing of ethanol impregnated tissue. Apparently, the ethanol impregnated tissue is transformed into an amorphous solid which, in turn, acts as a solid matrix supporting the delicate and highly ramified dendritic and axonal nerve cell processes. In addition, this solid matrix avoids, to a certain extent, the mechanical deformation of nerve tissue during fracturing.

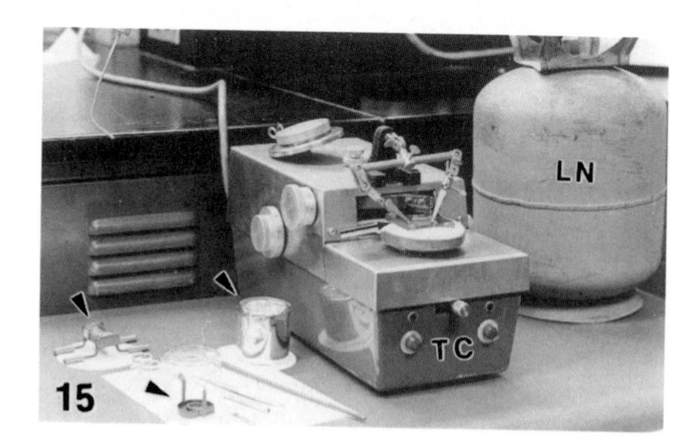

FIGURE 15. A modified Smith–Farquhar tissue chopper (TC) equipped with a liquid nitrogen copper stage (arrow) and a precooled fracture blade (arrow). The small cylinder contains liquid nitrogen (LN). Arrowheads indicate the accessories for storage and transferring the tissue samples.

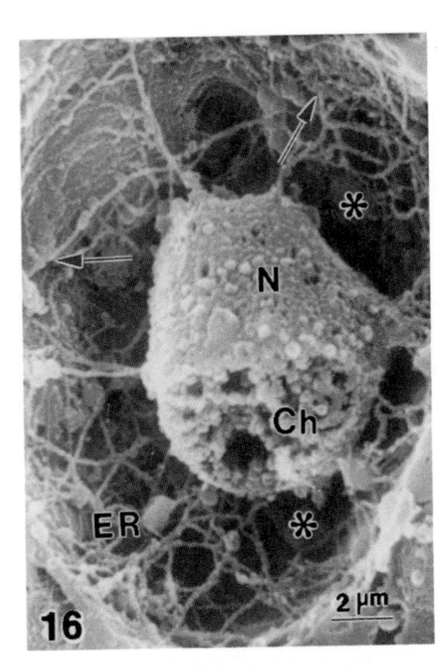

FIGURE 16. Fractured granule cell showing the nuclear chromatin (Ch), the outer surface of endoplasmic reticulum canaliculi (ER), and cisternae between the nucleus (N) and the inner surface of the plasma membrane (arrow). The freeze–fracture method for scanning electron microscopy "washes out" the cytosol, and causes cytoplasm anfractuous cavities (asterisks), that facilitate the cytomembrane visualization.

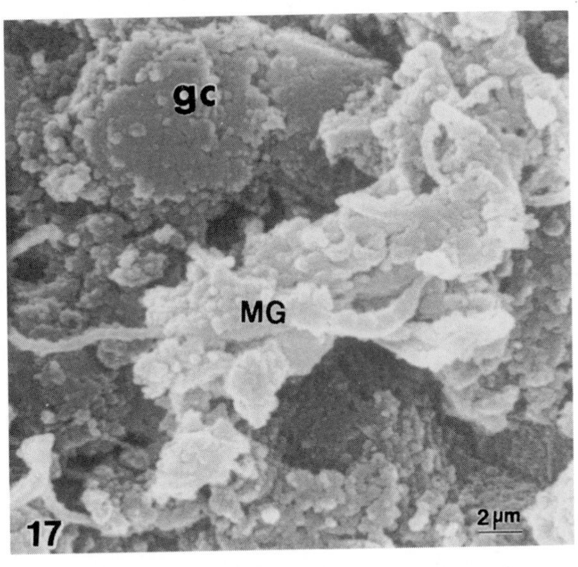

FIGURE 17. Granular layer of human cerebellar cortex. The ethanol-cryofracturing technique exposes the outer surface of granule cells (gc) and the cryo-dissected constitutive elements of mossy glomerular region (MG).

The limitations of this technique are as follows: (1) satellite neuroglial cells are easily removed during the cryofracture process; (2) slow freezing rate of liquid nitrogen; (3) coating with thick gold–palladium obscures nerve cell surface morphology (the gold–palladium masks the true nerve cell surface and obscures subsurface details); and

(4) the cryofracture process produces a random exposure of neuronal surface and sections of neuronal processes, thereby limiting the study of synaptic morphology.

Advantages are that the ethanol-cryofracturing technique is useful for tracing short intracortical circuits as well as for following the course of axonal collaterals until their terminus on a dendritic surface. In this context, the ethanol-cryofracture technique offers some advantages over the Golgi LM technique for studying neuronal geometry and nerve cell circuits. Specifically these advantages are: (1) better resolving power, (2) increased depth of focus, (3) continuity of nerve cell circuits, and (4) faithful delineation of three-dimensional pre- and postsynaptic structures. Both techniques used in conjunction allow a more accurate study of intracortical circuits (Castejón, 1988; Castejón & Castejón, 1988). The cryofracture technique of Humphreys et al. (1974, 1975) has been applied to the study of neural crest in the axolotl (Löfberg, Ahlfors, & Fallstrom, 1980), and to the human cerebellar cortex (Castejón, 1988; Castejón & Valero, 1980).

Freeze–Fracture SEM Method (Haggis & Phipps-Todd, 1977)

This method was applied to the study of teleost fish and primate central nervous system (Castejón, 1981, 1984, 1988, 1991). After perfusion fixation, cerebellar slices, 2–3 mm thick, were cut with a razor blade and fixed by immersion in the same fixative for 4–5 hr. After washing in buffered saline, they were postfixed in 1% OsO_4 in 0.1M phosphate buffer solution, pH 7.4, for 1 hr. After rinsing in a similar buffer, tissue blocks were dehydrated through graded concentrations of ethanol, rapidly frozen by plunging into Freon 22, cooled by liquid nitrogen (Haggis, 1970; Haggis & Phipps-Todd, 1977), and fractured with a precooled razor blade. The fracture fragments were returned to fresh absolute ethanol for thawing. Haggis and Phipps-Todd (1977) reported that the fix–Dehydrate–Freeze–fracture (fDFf) method, "washed out" the cytoplasmic and nuclear soluble proteins during the thawing step, thus leaving anfractuous cavities around the cytomembranes and allowing visualization of the surface details of cytoplasmic and nuclear structures (Figure 16). The tissue is then critical point dried with liquid CO_2, as recommended by Anderson (1951) and coated with gold–palladium. Specimens were examined in a JEOL 100B EM, with an ASID high resolution scanning device, at 10, 20, or 60 kV.

An interesting contribution of the SEM freeze–fracture method is that it provides a three-dimensional visualization of the spatial arrangement and organization of endoplasmic reticulum surface. In general, fractured cerebellar nerve cells showed a caveolar appearance that could be correlated with freezing and intracellular ice formation. Hunt, Taylor, and Pegg (1982) has also reported ice cavities and shrinkage of the perinuclear region in smooth muscle, an ideal model system for the investigation of cryo-injury. Therefore, the cytoplasmic damage observed in some cerebellar neurons probably are due to spherulitic crystallite formation (Luget, 1970). Apparently, a center of growing ice crystallization is formed in the cytoplasmic matrix, which repels the endoplasmic reticulum and cell organelles. Another alternative is cytoplasmic injury produced during rewarming and thawing by a process of devitrification, melting, or recrystallization. Since fragments of cerebellar tissue were plunged into Freon at −150°C (rapid freezing rate), there is the likelihood of intracellular freezing with cooling velocity or damage from recrystallization of intracellular ice during warming. However, in our experiments, water in the tissue was replaced with absolute ethanol prior to freezing, thus preventing ice crystal formation. The works of Humphreys et al. (1975) on liver, and Haggis, Bond, and Phipps-Todd (1976) on muscle suggest that in fixed tissues, ethanol freezes to a glass or noncrystalline state, both at rapid freezing rates (tissue plunged into Freon at −150°C) or medium freezing rates (tissue plunged directly into liquid nitrogen), with no distortion of tissue structure. However, in tables of physical constants, the melting point of ethanol is given as −112°C to −117°C; so, it is possible that ethanol may crystallize below these temperatures, thereby producing some structural rearrangement similar to that induced by ice formation. The freeze–fracture method also has been applied to the study of rat pineal body (Krstic, 1974) and confirmed communication between the pineal canaliculi and the pericapillary spaces.

Improved Freeze–Fracture SEM Method, Delicate Specimen Preparation, and Chromium Coating (Apkarian & Curtis, 1986; Castejón & Apkarian, 1992)

Excised Rhesus monkey cerebellar cortex is minced into 2 mm^2 pieces and further fixed in 2.5% glutaraldehyde in 0.1M cacodylate buffer, pH 7.4, overnight, in order to provide complete intracellular and extracellular proteinaceous cross-linking. Cacodylate buffer at pH 7.4, is used to completely remove the primary fixative by rinsing the tissue several times with gentle agitation. Postfixation of phospholipid moieties is accomplished by immersion in 1% OsO_4 in 0.1M cacodylate buffer, pH 7.4, for 1 hr and then rinsing in cacodylate buffer several times.

Delicate Specimen Preparation

A graded series of ethanol (30, 50, 70, 80, 90, 2 × 100%) is used to substitute tissue fluids, prior to wrapping individual pieces in preformed absolute ethanol filled parafilm cryofracture packets. Rapid freezing of packets is performed by plunging into Freon 22 at its melting point (−155°C) and then storing in liquid nitrogen. A modified tissue chopper (Sorval TC 2) equipped with a liquid nitrogen copper stage and a precooled fracture blade (−196°C) is utilized for cryofracture (Figure 15). First, the packet is transferred from the liquid nitrogen trough using chilled forceps in order to avoid thermal transfer. Secondly, the cooled fracture blade is removed from liquid nitrogen, the packet is orientated under the blade and the arm is immediately activated to strike only the top of the packet (Apkarian & Curtis, 1986). Fractured tissue fragments are transferred into chilled absolute ethanol (4°C) and thawed. Tissues are loaded into a fresh absolute ethanol filled mesh basket within the boat of a Polaron E-3000 critical point dryer and the boat loaded into the dryer and exchanged with CO_2 gas at a rate of 1.2 L/min. Then the CPD chamber is thermally regulated to the critical temperature and pressure at a rate of 1°C/min. Following the phase transition, the CO_2 gas is released at a gas flow rate of 1.2 L/min (Peters, 1980). Dried specimens, shiny face up, are mounted onto aluminum stubs 9 mm × 2 mm for the ISI (Topcon) DS-130 SEM upper stage, or onto brass mounts for the Hitachi S-900 SEM with silver paste and degassed at 5×10^{-7} Torr prior to coating.

Chromium Coating

Dried and mounted specimens are chromium coated with a continuous 1–2 nm film in a Denton DV-602 turbo pumped sputter deposition system operated in an argon atmosphere of 5×10^{-3} Torr (Apkarian & Joy, 1988).

Specimens are introduced onto the condenser/objective lens stage (SE-I signal mode operation) of either ISI DS-130 SEM equipped with LaB_6 emitter (Figure 18) or a Hitachi S-900 SEM equipped with a cold cathode field emitter. Both instruments are operated at 25–30 kV in order to produce minimal spot size and adequate signal to noise ratio at all magnifications. Micrographs are soft focus printed to reduce instrument noise (Peters, 1985a). This method provides an exploration into the inner and outer surfaces of vertebrate cerebellar neurons utilizing type I and II SEs (SE-I and SE-II) and topographic contrast (Castejón & Apkarian, 1992).

The outer nerve cell surface observations of unfractured nerve cells provide information on the three-dimensional morphology and surface appearance of nerve cell bodies. In order to study the outer surface of cerebellar nerve cells, granule and Golgi cells in teleost fishes and Rhesus monkeys were selected as models of cerebellar micro- and macroneurons because they are easily identified in the cerebellar granular layer and also because they present a large area for scanning the exposed somatic outer surface. Figure 14 illustrates Rhesus monkey cerebellar granular neurons. The 2-nm chromium coated image offers a smooth, brilliant, cotton-like appearance of nerve cell outer surfaces. This image can be compared with the opaque, finely granular, rough surface of the teleost fish nerve cells coated with 5–10 nm thick gold–palladium as illustrated in Figure 1.

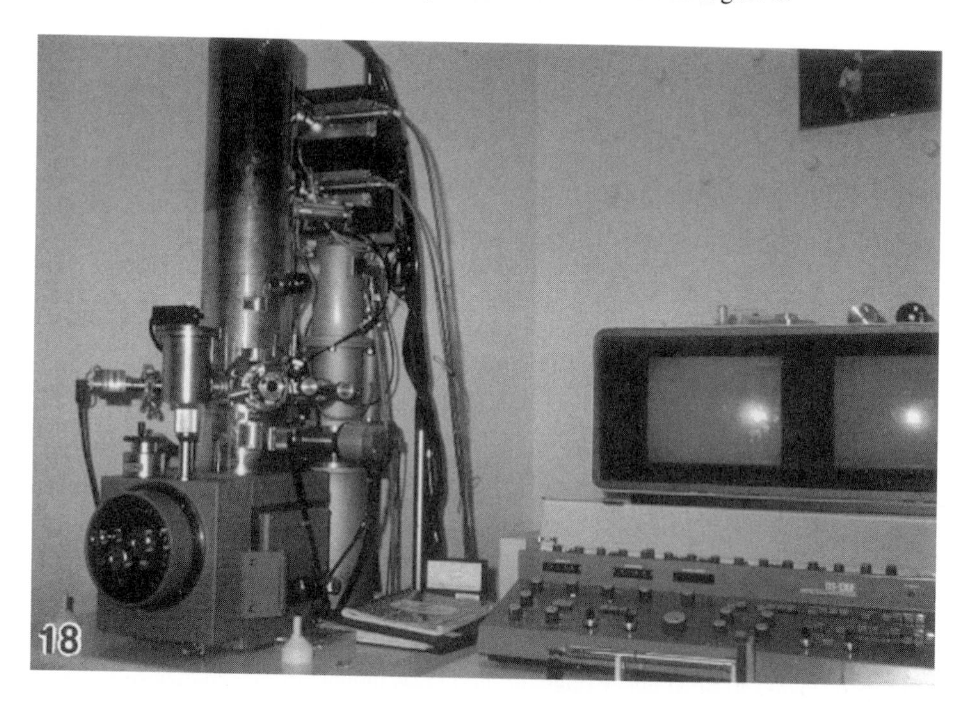

FIGURE 18. Scanning electron microscope equipped with a LaB_6 emitter.

The molecular layer of Rhesus monkey cerebellar cortex was studied at higher magnification in order to address the question of whether chromium coating allows the visualization of the extracellular space which is seen in ultrathin sections or direct replicas of TEM. Figure 19 illustrates the cryofractured molecular layer of Rhesus monkey cerebellar cortex showing the cross-section of parallel fibers limited by their SE-I profiles of plasma membranes and separated by the less dense, narrow extracellular space located between adjacent profiles. For comparative purpose, Figure 20 shows a corresponding region of an ultrathin section of mouse cerebellar cortex.

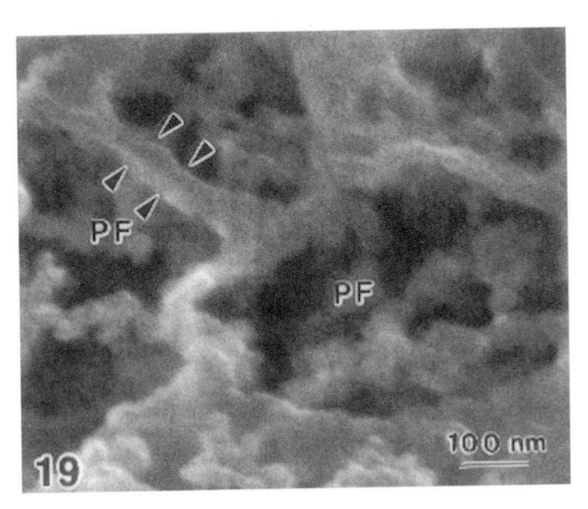

FIGURE 19. Fractured primate cerebellar molecular layer showing the type I secondary electron profile of limiting plasma membranes (arrowheads) of two cross-fractured parallel fibers (PF). The plasma membranes appear separated by a less dense extracellular space.

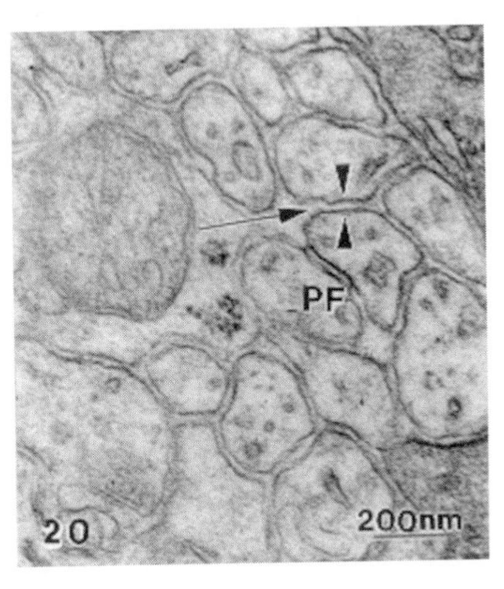

FIGURE 20. Corresponding TEM image of an ultrathin section of cerebellar molecular layer showing the cross sections of parallel fibers (PF), their limiting membrane (arrowheads), and the intervening extracellular space (arrow). Glutaraldehyde-osmium fixed cerebellum. Uranyl acetate and lead hydroxide stainings.

The freeze–fracture method allows examination of the outer somatic plasma membrane and intracellular membranes of nerve cells. At the thawing stage, the "wash out" of soluble proteins from the fracture face reveals the stereo-spatial relationship between cytoplasmic and nucleoplasmic compartments. At higher magnification and resolution, internal membranes, cytoskeleton, and chromatin arrangement can be studied in three-dimensional views. Variation in fixation conditions can be exploited to reveal different ultrastructural features. Slightly hypertonic fixative causes some cell shrinkage exposing the outer surface of the plasma membrane for SEM viewing. Fixation in hypotonic fixative induces cell breakage during fixation, swelling of cell organelles, and clear visualization of internal membranes (Haggis & Phipps-Todd, 1977).

Ultrasonic Microdissection (Arnett & Low, 1985; Low, 1989)

Rat cerebellar samples are initially fixed with aldehydes. Blocks of tissue are razor cut and treated by either of the three separate methods: (1) immersion in 1% aqueous boric acid; or (2) in 2% phosphate buffered OsO_4, followed by boric acid; or (3) in an 8 : 2 mixture of boric acid and OsO_4. After 18–48 hr of immersion, the blocks are dehydrated and exposed to ultrasound in 100% acetone, at frequencies of 80 kHz for 10–20 min. Then specimens are critical point dried using liquid CO_2, mounted on SEM stubs, and sputter-coated with gold–palladium. Microdissection of cut surfaces (erosion) occurs after all three treatments. All cerebellar cell types as well as neuronal processes and synaptic relationships in the granular layer are distinguished with this technique.

These techniques, when applied to cerebellum and spinal cord, also reveal, naturally occurring surfaces hitherto hidden from direct examination of their surface contours. Two responses of central nervous tissue became evident: (1) cavitation appears in the neuropil surrounding cell bodies when ultrasound is applied; and (2) the degree of microdissection depends on the concentration of aldehydes used in the initial cardiac perfusion of the experimental animal. Cavitation makes it possible to identify nerve cells by means of their position and general morphology.

The final contour of sonicated preparations, however, depends largely on procedures followed prior to exposure to ultrasound. The three separate methods of treatment used prior to sonication (boric acid, OsO_4 followed by boric acid and a mixture of both) were shown to produce distinctly different patterns of microdissection. Boric acid produces the least erosion of the cut surface, OsO_4 causes medium erosion, while OsO_4 followed by boric acid, somewhat more, and an 8:2 mixture of the two, the greatest erosion.

The Creative Tearing Technique (Scheibel, Paul, & Fried, 1981)

The cerebellar neuropil of the Mongolian gerbil, rat, cat, monkey, and human is revealed by careful tearing of aldehyde-fixed tissue specimens prepared by means of dehydration, CPD, and sputter-coating with gold–palladium. Scheibel et al. (1981) applied a modified free hand dissection to the tissue. After determining the desired plane of cleavage, a small cut with a scalpel or razor blade is made in a corner of the tissue block. Using the incision as the starting point, the tissue is slowly turned in the plane already established. The method follows the natural cleavage planes within the tissue. The technique provides a remarkable three-dimensional view of neuropil including cell bodies and dendrites, details of pre- and postsynaptic morphology, axonal structure, neuroglia, and the microvasculature.

In addition, the same research group (Paul, Fried, & Scheibel, 1984) has applied the "creative tearing" technique to another highly laminated structure, the hippocampus. Following intraventricular cardiac perfusion of the anesthetized animal with standard EM fixation media or, in the case of the human tissue, immersion of the tissue immediately into the fixative, the material is immediately dissected into grossly identifiable brain regions. These blocks remain in EM fixative for approximately 75 min. They are then dissected by manual tearing to reveal the desired surfaces, with care taken to avoid exposure of the tissue to air. The blocks are carefully trimmed under a dissecting microscope to a size of about 3 mm^3 and sketches made to preserve the proper orientation. Pieces are left overnight (about 14 hr) in refrigerated fixative. The following day, they are immersed in a 0.33% OsO_4 solution, for 1 hr. Dehydration is performed in a series of alcohols from 30% to 100%, with two 5-min changes in each, and three changes in the final 100% solution.

The dehydrated blocks are subjected to CPD using a Samdri PVT-3 CO_2 dryer, which takes the specimen from about 1300 psi (67,200 Torr) and 37°C down to atmospheric pressure and temperature. Since orientation is critical for ultimate interpretation of views obtained with the SEM, each block is placed in a separate well in the CPD apparatus, and further sketches are made, if necessary. The color of tissue after CPD is quite uniform, and accurate identification of regions can be difficult without a sketch. The pieces are removed immediately after the procedure and mounted on metal stubs or a thin strip of double-sided sticky tape. Colloidal graphite in isopropyl alcohol (DAG) is spread liberally on the stub and around the specimen to allow as much conduction surface as possible between tissue and metal surface. Since heat from the electron beam can melt the DAG, the stub is allowed to dry for at least 24 hr before viewing, causing movement of an incompletely dry specimen. Immediately before viewing, the specimen is sputter coated using gold–palladium in an argon atmosphere, for 2 min.

Paul et al. (1984) examined the gross morphology of hippocampal tissue, demonstrating several important aspects of both the dissection procedure and the three-dimensional relationship of the region. With this technique, the three-dimensional morphology of pyramidal cells was obtained displaying axons, dendrites, and dendritic spines.

Specimen Preparation Using a t-Butyl Alcohol Freeze-Drying Device (Hojo, 1996, 1998)

Hojo used 10% formalin-fixed human cerebellar cortex specimens, postfixed in 2.5% glutaraldehyde in 0.1M cacodylate buffer solution, and rinsed three times in 5% sucrose solution in the same buffer. After dehydration in a graded series of ethanol, the specimens are transferred into a graded series of t-butyl alcohol. The t-butyl alcohol substituted specimens are then freeze-dried at high vacuum (5×10^{-2} Torr). The specimens are finally sputter coated with gold. The scanning electron micrographs show good preservation of cellular organization of the cerebellar cortex and intact cytoarchitectonic arrangement. Granule and Golgi cells as well as Purkinje cells and their synaptic connections are clearly distinguished.

CONCLUDING REMARKS

Conventional and specialized nerve sample preparation techniques offer the unique possibility of displaying in three-dimensions the remarkable complexity of nerve cell organization and in situ interrelationship. Progress in the development and application of these preparative techniques to the study of different areas of the central and peripheral nervous system has resulted in a marked increase of our knowledge on neuronal geometry and outer and inner surface morphology. In addition, specialized techniques such as ethanol-cryofracturing, ultrasonic microdissection, and "creative tearing" have facilitated the study of nerve cell hidden surfaces or synaptic junctions, not readily accessible. Artifacts, however, are still present and further refinement of the preparation techniques is needed in order to obtain optimal preservation of nerve tissue. It is necessary to design specific preparative procedures tailored to the neurobiology of different areas of the central nervous system. This is especially important when considering the neurotransmitter heterogeneity of motor nervous centers compared with the autonomic nervous system.

Chapter 2

The Cerebellar White Matter

BRIEF HISTORY

The afferent and efferent fibers of cerebellar cortex were first described by Golgi (1874, 1882), Ramón y Cajal (1888), Kölliker (1890), Van Gehuchten (1891), Retzius (1892a), Lugaro (1894), Dogiel (1896), Held (1897), Athias (1897), and Bielschowsky and Wolff (1904). For a detailed review, see Ramón y Cajal (1911). In his monumental treatise, Ramón y Cajal described the connections of the cerebellum with other zones of central nervous system via the cerebellar peduncles. Eccles, Ito, and Szentágothai (1967), Brodal (1969), and Larsell and Jansen (1972) later reported detailed descriptions of afferent and efferent pathways. Elegant light microscopy (LM) and electron microscopic (EM) descriptions of afferent and efferent pathways were subsequently published by Palay and Chan-Palay (1974), and Ito (1984). Castejón and Sims (1999, 2000), Castejón, Castejón, & Sims (2000a), and Castejón, Castejón, and Alvarado (2000b) reported the confocal laser scanning microscopy (CLSM) and scanning electron microscopic (SEM) characterization of mossy and climbing fibers in the white matter of vertebrate cerebellar cortex.

THE AFFERENT AND EFFERENT FIBERS

The afferent and efferent fibers of the cerebellum are localized in the cerebellar peduncles that connect the cerebellar cortex with other components of central nervous system. The inferior peduncle contains the vestibular afferents, spinal afferents, lateral reticular nucleus and reticular formation which are some sources of mossy fibers. The afferents from the inferior olive represent the climbing fibers. The efferent fibers originate mainly from axons of Purkinje cells and from neurons in the fastigial nucleus.

The middle cerebellar peduncle contains mainly mossy fiber afferents from the pontine nuclei. The superior cerebellar peduncle contains afferent mossy fibers from spinal afferents and spinocerebellar tract. The efferent fibers originate from the interpositus, dentate, and fastigial nuclei. This book is mainly restricted to an examination of the cerebellar cortex by means of SEM; therefore, we will describe only the fibers contained in the white matter of the cerebellar folia. The reader is referred to the elegant book of Palay and Chan-Palay (1974) for a detailed description of the fiber connections and design of the cerebellar cortex.

At the center of each cerebellar folium lies a thin layer of white matter composed of myelinated afferent and efferent fibers, connecting the cerebellar cortex with other central nervous system zones. By means of low magnification SEM, the mossy and climbing fibers can be identified according to differences in their caliber and branching pattern.

Study of teleost fish cerebellar white matter with the scanning electron probe at low magnification of samples coated with gold–palladium shows longitudinal bundles of thick mossy parent fibers intermingled with groups of thin afferent climbing fibers (Figure 21). At higher magnification, both types of afferent fibers can be clearly distinguished by their differing thickness. The mossy fibers are up to 2.5 μm in diameter and the climbing fibers up to 1 μm in diameter, measured in cross sections of these fibers in conventional SEM fractographs. At the level of the entrance site to the granular layer, the afferent mossy and climbing fibers are additionally distinguished by their branching pattern (Figure 22). The mossy fibers exhibit a characteristic dichotomous pattern of bifurcation, whereas climbing fibers display a typical arborescence or crossing-over type of bifurcation. These criteria for identification at SEM level are in agreement with previous Golgi LM studies (Mugnaini, 1972). The cross-over which follows climbing fiber branching was first described by Athias (1897) and later by O'Leary, Inukai, and Smith (1971). The observations at SEM level were reported earlier by us (Castejón, 1983b) and have been later supported by correlative CLSM and SEM examinations (Castejón et al., 2000a, 2000b). The only efferent fibers from the cerebellar cortex are represented by the Purkinje cell axons. These

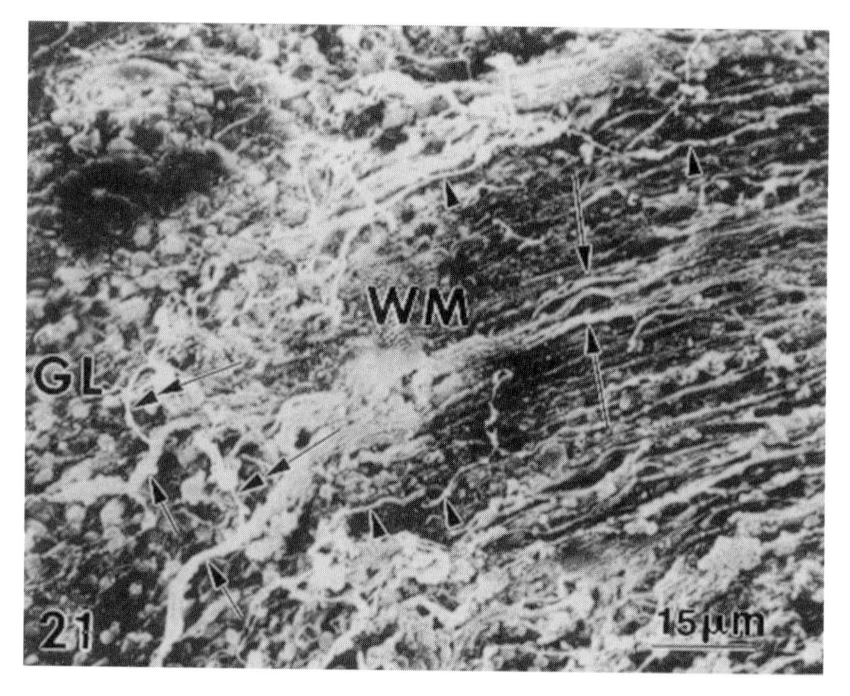

FIGURE 21. Teleost fish cerebellar cortex. White matter (WM) of cerebellar folia showing thick parent mossy fibers (long arrows) and thin climbing fibers (arrowheads) entering the granular layer (GL). The thickest fibers correspond to Purkinje cell axons (short arrows) and exhibit recurrent collateral processes (doublehead arrows). Gold–palladium coating.

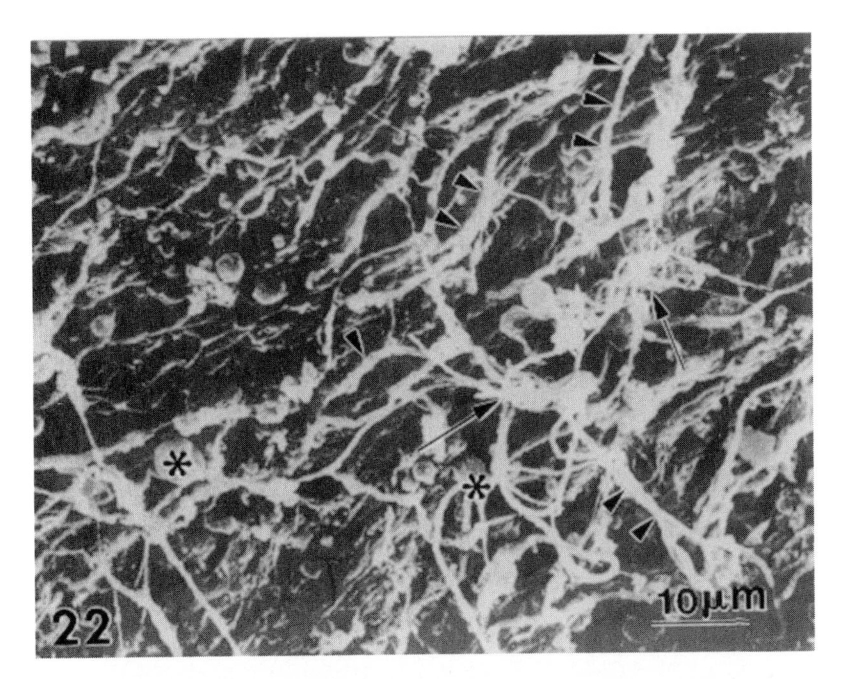

FIGURE 22. Teleost fish cerebellar white matter showing climbing fibers which exhibit the typical crossing-over branching pattern (arrows), featured by fine collaterals spreading in three different planes. The mossy fibers (arrowheads) show the characteristic dichotomous bifurcation pattern. Oligodendrocytes (asterisks) are also distinguished associated with the afferent fibers. Gold–palladium coating.

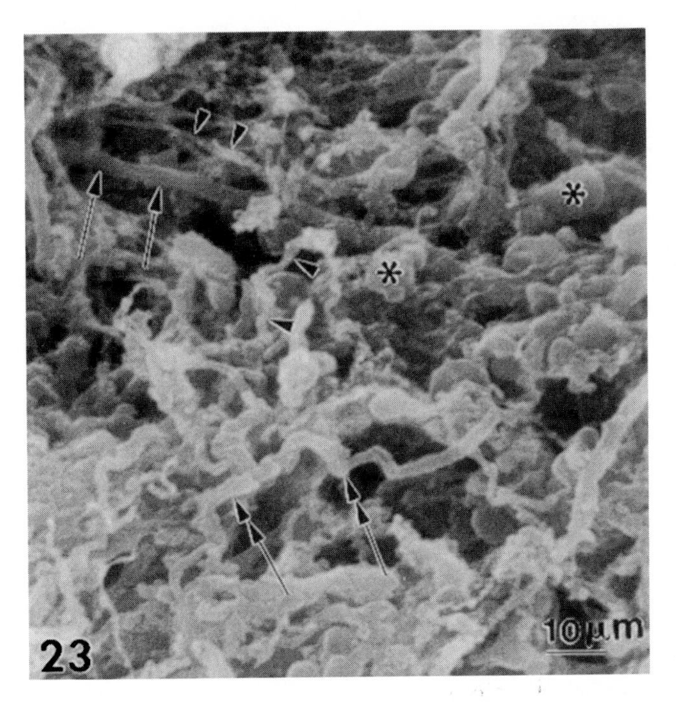

FIGURE 23. Teleost fish cerebellar white matter showing a thick Purkinje cell axon (arrows) with recurrent collateral processes (arrowheads). Another axon exhibits a twisted course (doublehead arrows) and varicosities presumably belonging to a monoaminergic fiber. Oligodendrocytes (asterisks) are also noted. Gold–palladium coating.

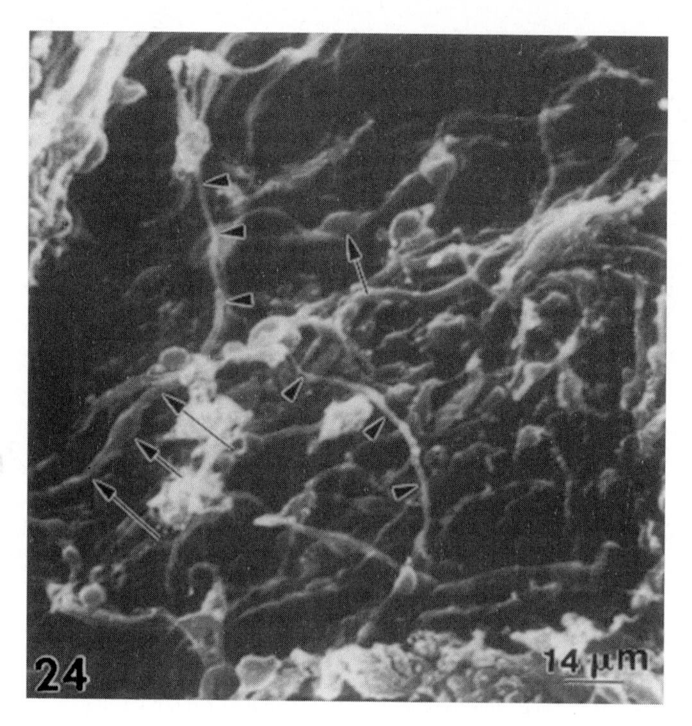

FIGURE 24. Teleost fish cerebellar white matter showing the twisted course of a Purkinje cell axon (long arrows), the recurrent collaterals (arrowheads), and their varicosities (short arrows). Gold–palladium coating.

axons originate from the lower pole of Purkinje cell soma in the Purkinje cell layer and proceed directly through the granular layer to enter into deep white matter. These axons are thicker than mossy and climbing fibers and exhibit varicosities and collateral processes (Figures 23 and 24) that return upward to the granular layer; hence, their name as recurrent collaterals (Ramón y Cajal, 1911, 1955). These collaterals are distributed in the granular and Purkinje cell layers and participate in the formation of Purkinje cell infraganglionic plexus. The third type of cerebellar afferents, the monoaminergic fibers (noradrenergic and serotoninergic), presently cannot be identified purely by SEM morphological features, but rather by immunocytochemical techniques (Abbott & Sotelo, 2000; Kerr & Bishop, 1991).

CONCLUDING REMARKS

The three-dimensional characterization of mossy and climbing fibers in the cerebellar white matter has been made by conventional SEM, in accordance with their thickness and branching pattern. The mossy fibers appear as thick parent fibers, up to 2.5 μm in diameter, with a characteristic dichotomous bifurcation pattern. The climbing fibers are thinner fibers, up to 1 μm in diameter, with a typical crossing-over bifurcation pattern. The efferent Purkinje cell axons appear as the thickest fibers, with recurrent collateral processes.

Chapter 3

Granule Cells

SHORT HISTORY

Golgi (1874, 1883) and Ramón y Cajal (1888, 1890a, 1890b) made the first light microscopic (LM) descriptions of cerebellar granule cells and their synaptic connections with afferent and intrinsic fibers of cerebellar cortex. Later, Estable (1923) and Fox and Barnard (1957) supported and extended such pioneering studies. Subsequently, Golgi LM and transmission electron microscopic (TEM) research reported additional morphological and submicroscopic features (Castejón, 1969; Dahl, Olsen, & Birch-Anderson, 1962; Eccles, Ito, & Szentágothai, 1967; Fox, 1962; Fox, Siegesmund, & Dutta, 1964; Fox, Hillman, Siegesmund, & Dutta, 1967a; Garcia-Segura & Perrelet, 1984; Gray, 1961; Hillman, 1969; Landis & Reese, 1974; Mugnaini, 1972; Palay & Chan-Palay, 1974; Sotelo, 1969).

By means of low temperature dry ashing technique, Lewis (1971) showed the first SEM images of the cat silver stained granule cells. Messer (1977) described postnatal mouse cerebellar granule cells, in monolayer culture. Using the slicing technique for conventional scanning electron microscopy (SEM), Castejón and Caraballo (1980a, 1980b), Castejón (1981), and Castejón and Castejón (1981) first described teleost fish granule cells. Castejón and Valero (1980) described the granule cells of human cerebellar cortex by employing the ethanol-cryofracturing technique of Humphreys, Spurlock, and Johnston (1974). Scheibel, Paul, and Fried (1981) exposed the granule cell outer surface using the creative tearing technique. Castejón (1981) applied the freeze–fracture method for SEM to teleost fish cerebella and described the inner cytoplasmic and nuclear organization of granule cells. Grovas and O'Shea (1984) analyzed by SEM the inward migration of external granular cells to form the internal granular layer. Sievers, Mangold, and Berry (1985) reported the ultrastructure, surface morphology, and synaptic connectivity of ectopic granule cells induced by 6-hydroxydopa (6-OHDA). Arnett and Low (1985) and Low (1989) performed ultrasonic microdissection of rat cerebellar cortex to show the granule cell outer surface. Castejón (1984) compared the three-dimensional morphology of granule cell soma by means of freeze-etching direct replica technique and freeze–fracture method for SEM. Later, Castejón (1988, 1990a, 1990c) described the parallel fiber–Purkinje dendritic spine synaptic relationship using freeze–fracture SEM and the freeze-etching technique. Apergis, Alexopoulos, Mpratokos, and Katsorchis (1991) characterized granule cells in rat cerebellar cortex. Castejón and Apkarian (1992, 1993) reported conventional and high resolution field emission scanning electron microscopy (FESEM) features of parallel fiber–Purkinje spine synapses. Castejón, Castejón, and Apkarian (1994a, 1994c) and Castejón, Apkarian, and Valero (1994b) studied high resolution SEM and proteoglycan ultracytochemistry of parallel fiber presynaptic endings. Hojo (1994) using the t-butyl alcohol freeze substitution device showed a sharp granule cell outer surface. Castejón (1996) described the organization of parallel fiber presynaptic endings, the type I secondary electron (SE-I) features of the synaptic membrane complex with Purkinje dendritic spines, and the visualization of postsynaptic receptors with high resolution FESEM. Castejón and Castejón (1997) reported the three-dimensional morphology by SEM of parallel fiber–Purkinje dendritic spine synapses. More recently, Castejón and Castejón (2000), Castejón, Castejón, and Sims (2000a, 2001a), Castejón, Castejón, and Alvarado (2000b), Castejón, Apkarian, Castejón, and Alvarado (2001b), and Castejón, Castejón, and Apkarian (2001c) performed comparative and correlative microscopic studies, to demonstrate the synaptic relationships of granule cell dendrites with mossy and climbing fibers and Golgi cell axonal ramifications.

THE THREE-LAYERED STRUCTURE OF CEREBELLAR CORTEX

The freeze–fracture method for SEM applied to teleost fish cerebellar cortex exposes the cytoarchitectonic arrangement of the cerebellar cortex (Figure 25). The granule cell, Purkinje cell, and molecular layers can be examined from low to high magnifications taking advantage of the depth of focus and the high illuminating electron probe of conventional SEM and FESEM. During cerebellar sample preparation, the cleavage plane of the cryofracture method is produced at the level of extracellular space existing among neuroglial cells and neurons exposing the outer surface of the latter cells. Most cerebellar neuroglial cells are partially and selectively removed during the cryofracture process facilitating the visualization of neuronal outer surfaces.

OUTER SURFACE OF INTACT GRANULE CELLS

One of the main contributions of SEM and the SEM cryofracture method is providing visibility of the outer surface view of granule cells. These observations on the outer surface view are essential for future studies using labelling techniques, detection of antigenic sites, and macromolecular assemblies of complex carbohydrates attached to the outer surface of neuronal plasma membrane or glycocalyx layer (Luft, 1971). In the granular layer of teleost fish cerebellum examined by low magnification SEM, the granule cell groups appear to be formed by 7–12 microneurons separated by the glomerular islands and surrounded by isolated Golgi cells. Examination by very low magnification SEM to localize the granular layer in the fragments of cerebellar folia is an obligatory step for proper orientation. As illustrated in Figure 26, an “en face” view of mouse granule cell groups is first observed in the fragment of cerebellar cortex (Figure 26). An almost complete view of the granular layer organization can be appreciated with low magnification FESEM (Figure 27). The depth of focus and low resoluting power of conventional SEM and FESEM allow simultaneous examination of the cytoarchitectonic arrangement of the granular layer and the participating granule cells in each group, and to trace the afferent mossy and climbing fibers with their synaptic connections.

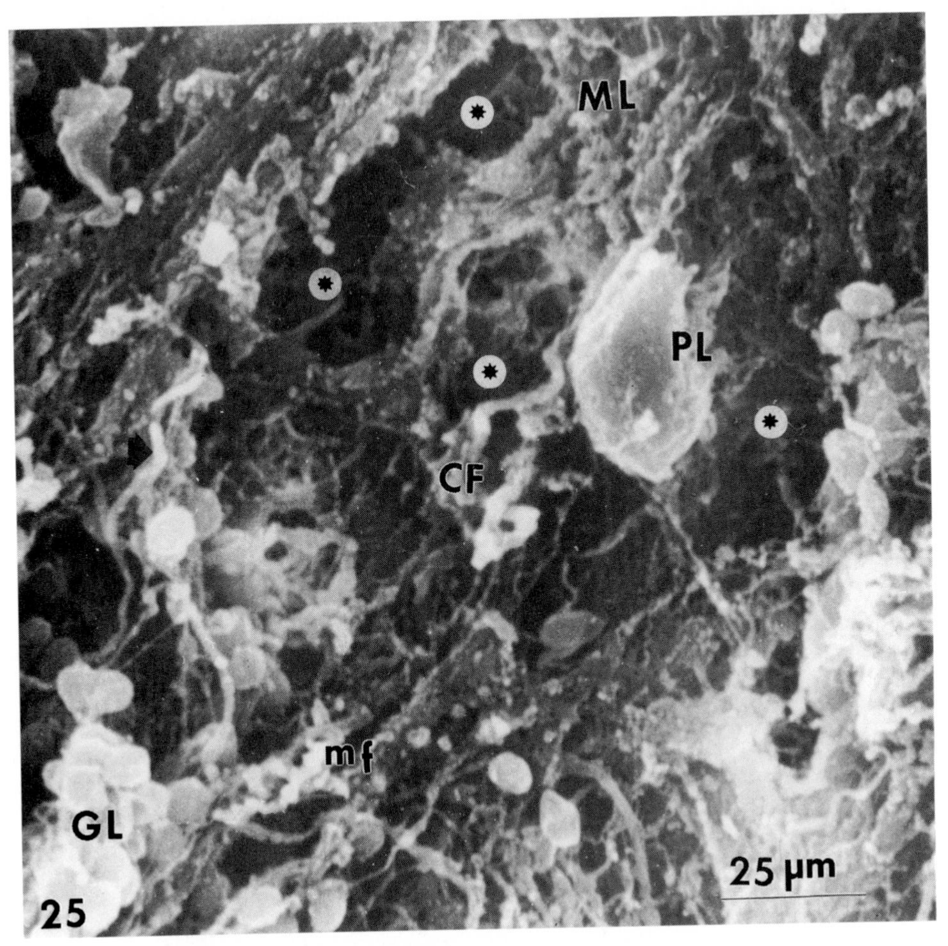

FIGURE 25. Low magnification scanning electron micrograph showing teleost fish cerebellar cortex following cryofracturing. The granular (GL), Purkinje (PL), and molecular (ML) layers are easily distinguished. Mossy fibers (mf) appear in close topographic relationship with granule cells. Climbing fiber (CF) can be seen approaching the Purkinje cell soma. The cryofracture process selectively removed the neuroglial cell layer, thus exposing the outer surface of cerebellar neurons. Asterisks label the dark spaces previously occupied by glial cells. Gold–palladium coating.

Close examination of granule cell geometry with the electron beam facilitates the visualization of granule cell soma and its axonal and dendritic processes (Figure 28). At higher magnification, the axon hillock region exhibits a triangular shape and the initial axonal segment can be seen as it crosses the granular layer and ascends to the molecular layer (Figure 29).

SEM images show wider spaces separating the granule cells. This fact reveals volume and dimensional changes occurring during SEM sample preparation. As previously mentioned the glutaraldehyde commonly used as the primary fixative generally causes shrinkage of nerve cells mainly because slightly hypertonic fixatives (320 mOsm) are used to obtain proper fixation of nerve tissue (Castejón, 1993).

Standard dehydration schedules inevitably cause cells and tissue to swell and shrink (Boyde, Bailey, Jones, & Tamarind, 1977), often causing cracks. In addition, shrinkage

FIGURE 26. Low magnification scanning electron micrograph of mouse cerebellar cortex. Slicing technique. Arrows indicate the sectioned end of the specimen. An "en face" view of granule cell groups (gc) is obtained at the upper side of specimens. Gold–palladium coating.

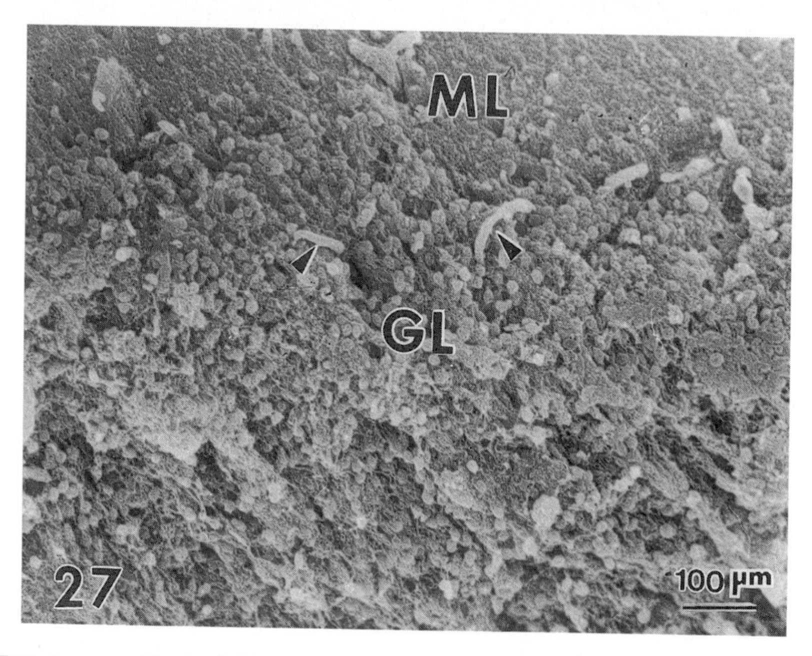

FIGURE 27. Low magnification field emission scanning electron micrograph of Rhesus monkey cerebellar cortex showing the granule cell groups at the level of the granular layer (GL). The molecular layer (ML) and capillaries (arrowheads) are also seen. Chromium coating.

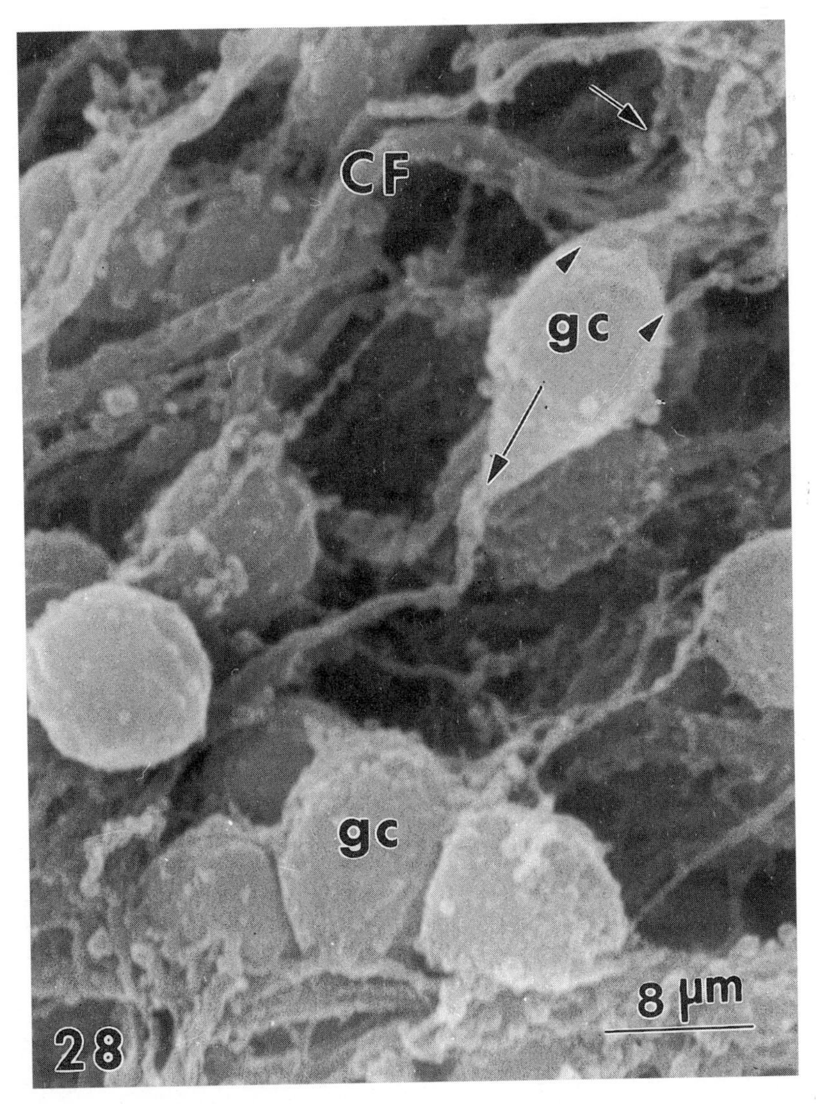

FIGURE 28. Conventional scanning electron micrograph of teleost fish cerebellar granular layer. The three-dimensional morphology of a granule cell (gc) and their axonal (long arrow) and dendritic processes (arrowheads) are seen. The outer surface of climbing fiber (CF) synaptic contacts (short arrow) with granule cell dendrites is also seen. Gold–palladium coating.

seems to be the main artifact caused by critical point drying (CPD) method. The cryofracture method as applied to the cerebellar cortex also should be considered. A 20% reduction in linear dimensions after this procedure may seem acceptable (Gamliel, 1985). Freeze–fracture SEM methods show that neighboring granule cells are close to one another at the granule cell groups, without intervening glial cytoplasm, and their plasma membranes are separated by a 20-nm extracellular space. This separation corresponds to the membrane-to-membrane space described in ultrathin sections by Palay and Chan-Palay (1974) and Castejón et al. (2001c).

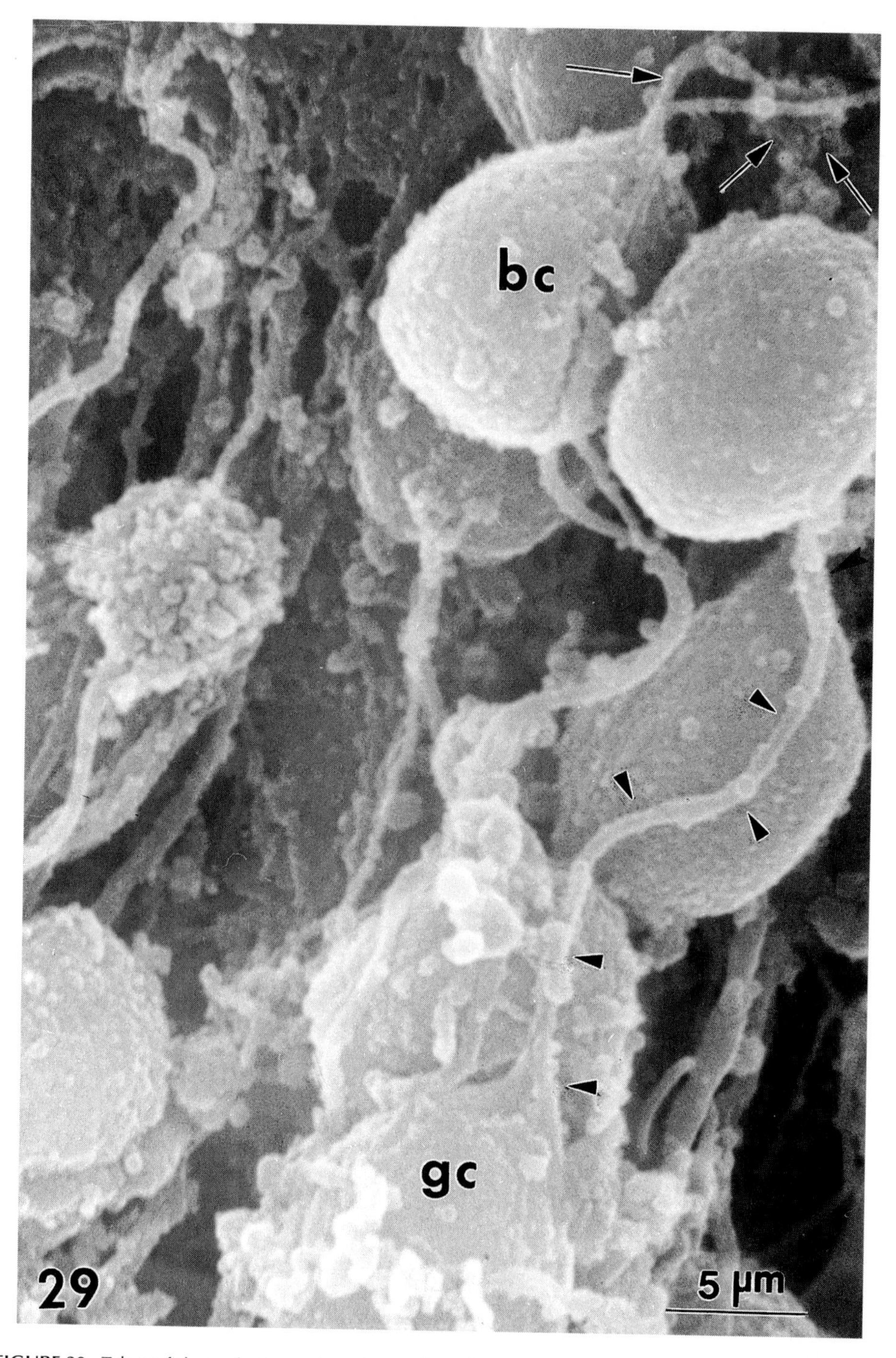

FIGURE 29. Teleost fish cerebellar cortex. A granule cell (gc) shows a triangular axon hillock region giving rise to its ascending axonal process (arrowheads). A unipolar brush cell (bc) is observed at the upper right side of the figure showing a dendritic process (long arrow) ending in a paintbrush structure (short arrows). Gold–palladium coating.

INNER ORGANIZATION OF FRACTURED GRANULE CELLS

The SEM freeze–fracture method (Haggis & Phipps-Todd, 1977), was applied to the *Salmo trout* cerebellum (Castejón & Caraballo, 1980b) in order to illustrate the closely packed granule cells (Figure 30), the plasma membranes of which appeared to be in intimate apposition. The fracture plane passes across the equatorial plane of cell somata allowing us to

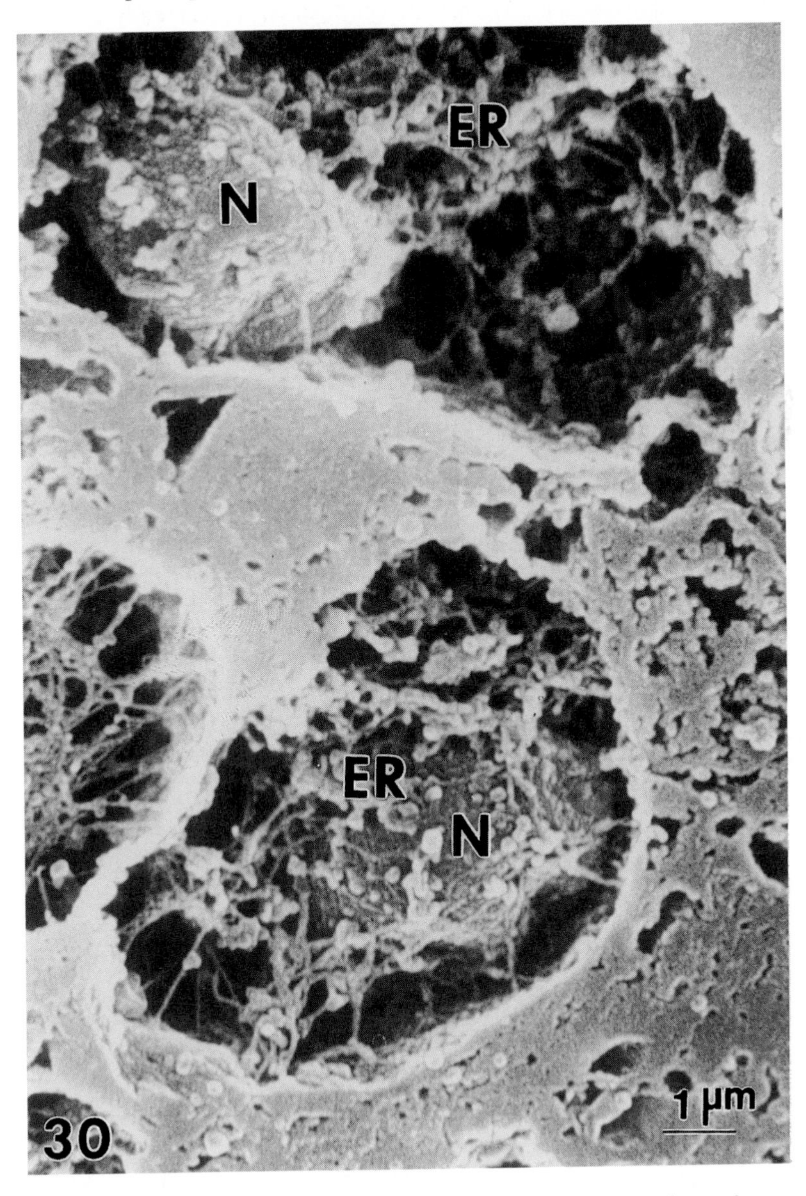

FIGURE 30. Teleost fish cerebellar cortex. The scanning electron microscopy freeze–fracture method shows two fractured granule cells exhibiting the nuclear (N) and endoplasmic reticulum (ER) outer surface. The cytoplasmic matrix has been washed out during the freeze–fracture process allowing visualization of inner surface structures. Gold–palladium coating.

visualize the outer surface of the nucleus that is held in a central or peripheral position by some cytoplasmic strands. These fine fibrils extend from the nuclear outer surface to the inner surface of the plasma membrane and form a cytoplasmic network where mitochondria and lysosomes are suspended. The arrangement and disposition of these cytoplasmic strands correspond to the distribution of granule cell endoplasmic reticulum. The cytoplasmic strands exhibit small or medium size varicosities or globular enlargements that correspond to the cisternae and vesicles of endoplasmic reticulum. The remaining cytoplasm appears nearly almost empty due to the fact that the freeze–fracture method "washes out" the soluble proteins of cytoplasmic matrix and exposes the three-dimensional view of the granule cell interior. This image has been correlated with the TEM appearance of granule cell morphology (Castejón, 1969, 1981, 1984; Castejón, Castejón, & Apkarian, 2001c; Mugnaini, 1972; Palay & Chan-Palay, 1974).

GRANULE CELL PROCESSES

With conventional SEM, the granule cells appear as smooth surface spheroid cells that give rise to a filiform axon, from a triangular shaped axon hillock (Figures 28 and 29). This axon crosses the interspace between the neighboring granule cell bodies and ascends to the molecular layer. The importance of this ascending granule cell axon has been recently emphasized by Gundappa-Sulur, De Schutter, and Bower (1999). After bifurcating in the molecular layer, granule cell axons or parallel fibers can be identified in the *Salmo trout* cerebellar cortex (Figure 31) as axonal bundles following a perpendicular pathway to the

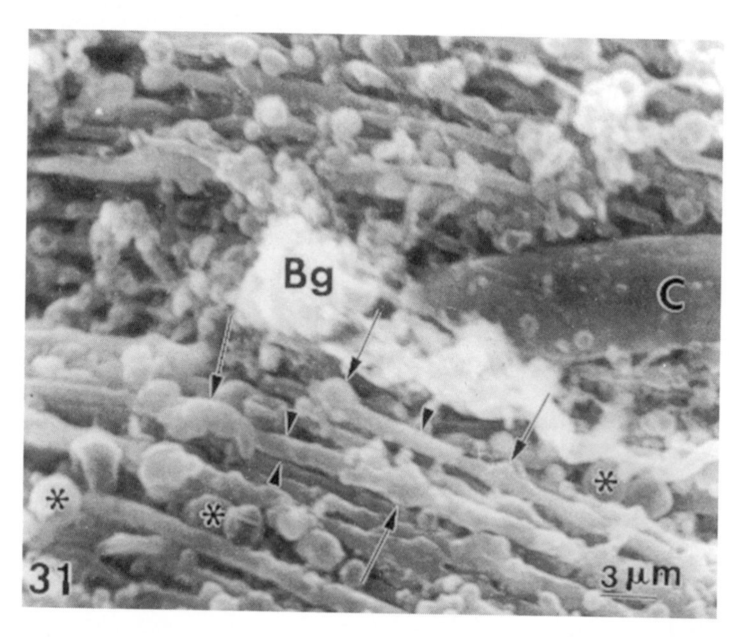

FIGURE 31. Teleost fish cerebellar cortex. Freeze–fracture scanning electron microscopy (SEM) method. Molecular layer showing the outer surface of granule cell axons or parallel fibers, which exhibit the synaptic varicosities (arrows) and nonsynaptic segments (arrowheads). Some vestiges of Bergmann cell cytoplasm (Bg) and a capillary (C) are also seen. Some globular structures (asterisks) correspond to Purkinje dendritic spine body. Gold–palladium coating.

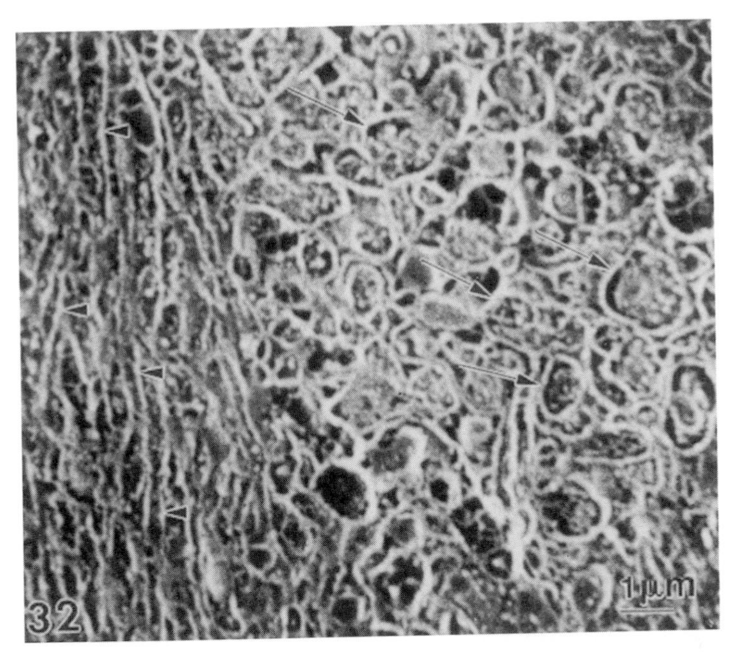

FIGURE 32. Teleost fish cerebellar cortex. Freeze–fracture scanning electron microscopy method. Molecular layer showing the longitudinally fractured nonsynaptic segments of parallel fibers (arrowheads) and the cross fractured synaptic varicosities (arrows) containing spheroidal synaptic vesicles. Gold–palladium coating.

Purkinje dendritic arborization, as classically shown by Golgi LM studies. At higher magnification, they exhibit a filiform shape and successive enlargements or synaptic varicosities; therefore, we have distinguished parallel fiber non-synaptic segments from *en passant* synaptic varicosities or synaptic segments (Figure 32). The cross-sections of both can be seen in the immediate neighborhood of the primary, secondary, and tertiary Purkinje dendrites. The *en passant* synaptic varicosities contain spheroidal synaptic vesicles. The nonsynaptic segments are observed as smaller, round profiles containing axonal microtubules.

CONCLUDING REMARKS

Conventional SEM and FESEM provide a three-dimensional image of granule cells and their topographic relationship with afferent climbing and mossy fibers. By means of the slicing technique and ethanol-cryofracturing technique, it is possible to study granule cell neuronal geometry. The depth of focus of SEM offers the additional possibility for studying granule cell participation in the cytoarchitectonic arrangement of the granular layer. The freeze–fracture method for SEM enabled us to visualize nuclear and cytoplasmic organization of fractured granule cells.

Chapter 4

The Mossy Fiber Glomerulus

THE LONG HISTORY

The cerebellar protoplasmic islands and the cerebellar glomeruli were studied at the light microscope level by Denissenko (1877), Ramón y Cajal (1890b, 1926), Retzius (1892a, 1892b), Lugaro (1894), Held (1897), Bielschowsky and Wolff (1904), Berliner (1905), Estable (1923), Cragie (1926), Jakob (1928), Pensa (1931), Boecke (1942), and Fox and Bertram (1954). Later, Brodal and Drablos (1963) distinguished two types of mossy fibers. Ramón y Cajal (1890b), using Golgi preparations, initially described the articulation between the mossy fibers and the granule cell dendrites at the level of the mossy glomeruli and also the participation of Golgi cell axonal ramifications. His earlier observations, later confirmed by the above-mentioned authors, led him to postulate the concept of a synapse by gearing. The views of Ramón y Cajal, as summarized in the Neuronal Doctrine (Ramón y Cajal, 1954), contrasted with the opinion of those supporting the reticularist theory, which conceived the continuity of nerve cell processes forming a diffuse neuronal network (Golgi, 1874; Pensa, 1931).

The cerebellar islands were among the first regions of nerve tissue analyzed with the advent of transmission electron microscopy (TEM). The pioneering studies of Palay (1956) and Gray (1961) confirmed, at the level of the cerebellar glomeruli, the validity of the Ramón y Cajal's neuronal theory. Subsequent TEM studies of the glomerular islands in several vertebrates established the asymmetric or Gray's type I synaptic contacts between the mossy fiber rosettes and the granule and Golgi cell dendritic digits (Castejón, 1971; Fox, 1962; Fox, Hillman, Siegesmund, & Dutta, 1967a; Hámori, 1964; Hámori & Szentágothai, 1966b; Mugnaini, 1972; Palay & Chan-Palay, 1974; Sotelo, 1969). According to these studies, three-dimensional diagrams of the synaptic connections were made by Eccles, Ito, and Szentágothai (1967) and Mugnaini, Atluri, and Hank (1974) in an attempt to provide a three-dimensional image of a multisynaptic complex. In our laboratory, and using the ethanol-cryofracturing technique (Castejón & Valero, 1980) we have described the synaptic relationship between mossy fibers and granule cell dendrites in human cerebellar cortex. In addition, we traced the course of mossy fibers through the granular layer in teleost fish cerebellar cortex prepared according to the slicing technique for conventional scanning electron microscopy (SEM) (Castejón, 1981). In such studies, the large depth of focus of SEM allowed us to characterize the mossy fibers and to compare the results with TEM studies. Scheibel, Paul, and Fried (1981) showed the dendritic and axonal structures of the gerbil cerebellar glomeruli using the "creative tearing" method. By means of an improved method of preparing central nervous system for examination at the SEM level, Reese, Landis, and Reese (1985) visualized the complex arrangement of neurites in the glomerulus. Arnett and Low (1985) exposed by ultrasonic microdissection, the granule cell clusters around the termination of mossy fibers and the glial elements involved in the formation of rat mossy glomeruli. As described by Palay and Chan-Palay (1974), we have tried to distinguish mossy fiber glomeruli from climbing fiber glomeruli in the cerebellar granular layer (Castejón, 1983b, 1986, 1988; Castejón & Castejón, 1988). Jakab and Hámori (1988) and Jakab (1989) reported a three-dimensional reconstruction, synaptic architecture and quantitative morphology of rat cerebellar glomeruli. Castejón and Castejón (1991) reported the three-dimensional morphology and proteoglycan content of mossy fiber glomeruli. More recently, Jaarsma, Diño, Cozzari, and Mugnaini (1996) described acetyltransferase positive mossy fibers with their granular and unipolar brush cell targets.

THE MOSSY FIBER–GRANULE CELL SYNAPTIC RELATIONSHIP

Using the slicing technique and gold–palladium coating, conventional SEM permitted the identification on teleost fish cerebellar cortex of the thick parent mossy fibers, 1–2 μm in diameter, and their collateral branches in the granular layer (Figure 33). They can be traced passing from one granule cell group to another configuring *en passant* synaptic

FIGURE 33. Teleost fish cerebellar cortex. Freeze–fracture scanning elctron microscopy (SEM) method. Granular layer. Two thick parent mossy fibers (mf) are observed running through the granular layer. Golgi (go) and granule cells (gc) are clearly distinguished by their different sizes. Numerous thin undifferentiated collateral processes (arrowheads) are seen. SEM slicing technique. Gold–palladium coating.

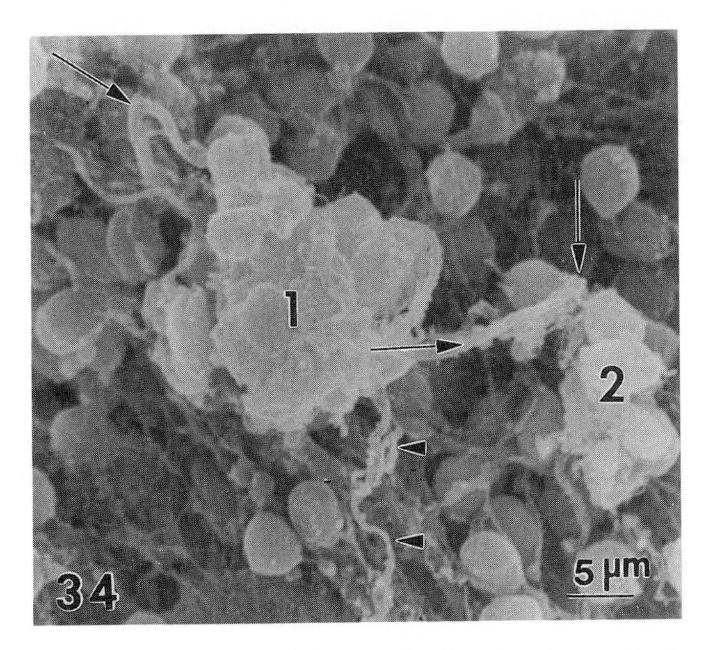

FIGURE 34. Teleost fish cerebellar cortex. A mossy fiber (arrow) is seen entering into two different granule cell groups (1,2) supporting the *en passant* nature of mossy fiber synaptic contacts. Fine axonal processes (arrowheads) presumably corresponding to Golgi cell axonal ramifications, also participate at the mossy glomerulus. Scanning electron microscopy (SEM) slicing technique. Gold–palladium coating.

contacts (Figure 34), supporting earlier Golgi light microscopic studies (Castejón & Caraballo, 1980b). By means of the ethanol-cryofracturing technique applied to human cerebellum, the outer surface of granule cell dendritic processes are observed establishing topographic contacts with afferent mossy fibers (Figure 35). Applying the freeze–fracture method for SEM (Castejón 1988; Castejón, Castejón, & Sims, 2000a), a sagittal fracture disclosed the longitudinal trajectory of mossy fibers and the formation of rosette expansions, which appear surrounded with the granule cell dendritic twigs (Figure 36).

An *en face* view of the glomerular region obtained by the SEM slicing technique shows up to 18 granule cells surrounding the mossy fiber rosette. The granule cell dendrites appeared to be converging radially toward the mossy rosette (Figure 37). These novel three-dimensional views of the glomerular region, allow for a more precise estimation of the quantitative relation of the degree of divergence of information of a mossy fiber upon groups of granule cells as well as the degree of convergence of many dendrites of different granule cells upon a central mossy fiber rosette.

CONCLUDING REMARKS

The above-described images emphasize the unsurpassed potential of SEM and the freeze–fracture method for studying the three-dimensional morphology of multisynaptic complexes in the central nervous system. Due to the depth of field of SEM, the dendritic

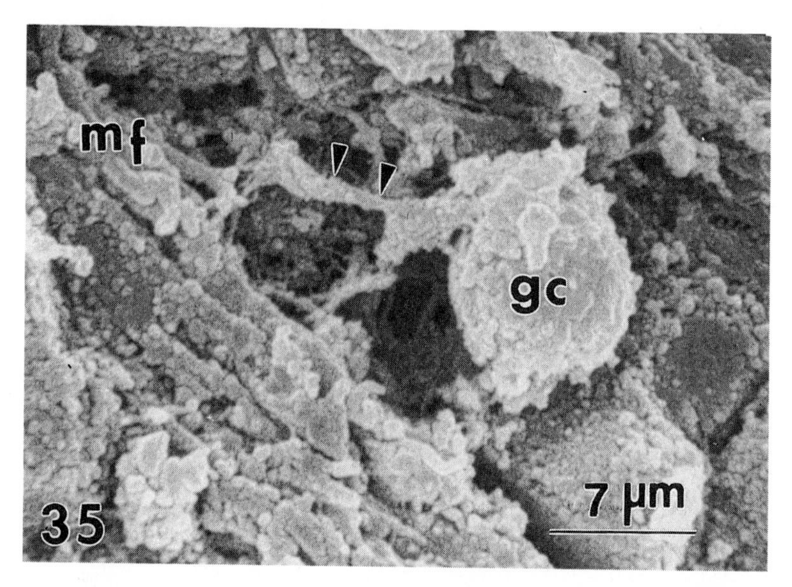

FIGURE 35. Human cerebellar cortex. Granule cell (gc) showing a dendritic process (arrowheads) in topographic contact with a mossy fiber (mf). Ethanol-cryofracturing technique. Gold–palladium coating.

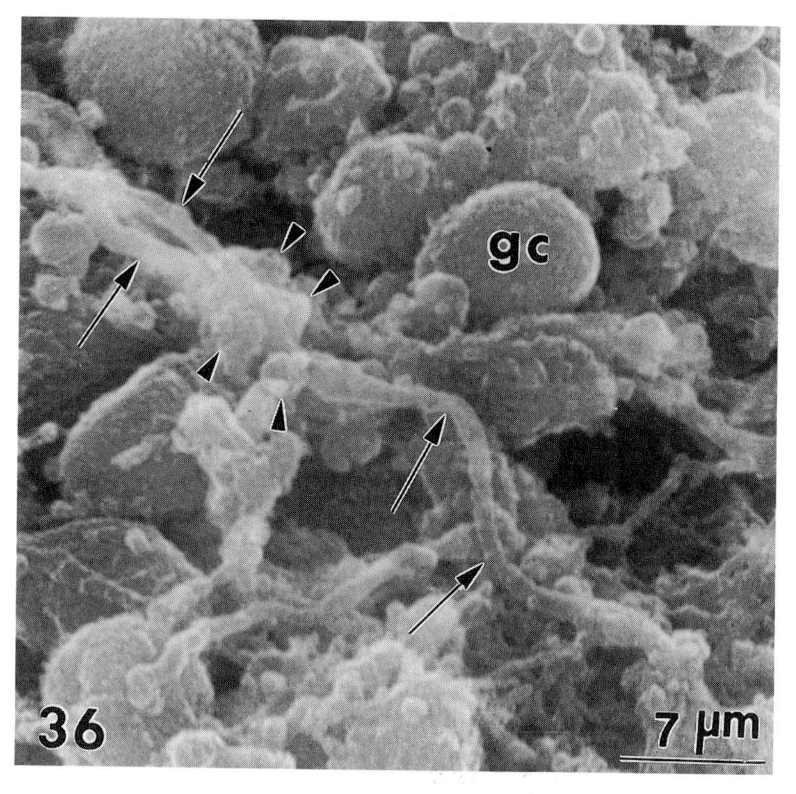

FIGURE 36. Teleost fish cerebellar cortex. Sagittally fractured mossy fiber glomerulus showing the longitudinal profile of a central mossy fiber (arrows) surrounded with the dendritic tips (arrowheads) of granule cells (gc). Scanning electron microscopy (SEM) freeze–fracture method. Gold–palladium coating.

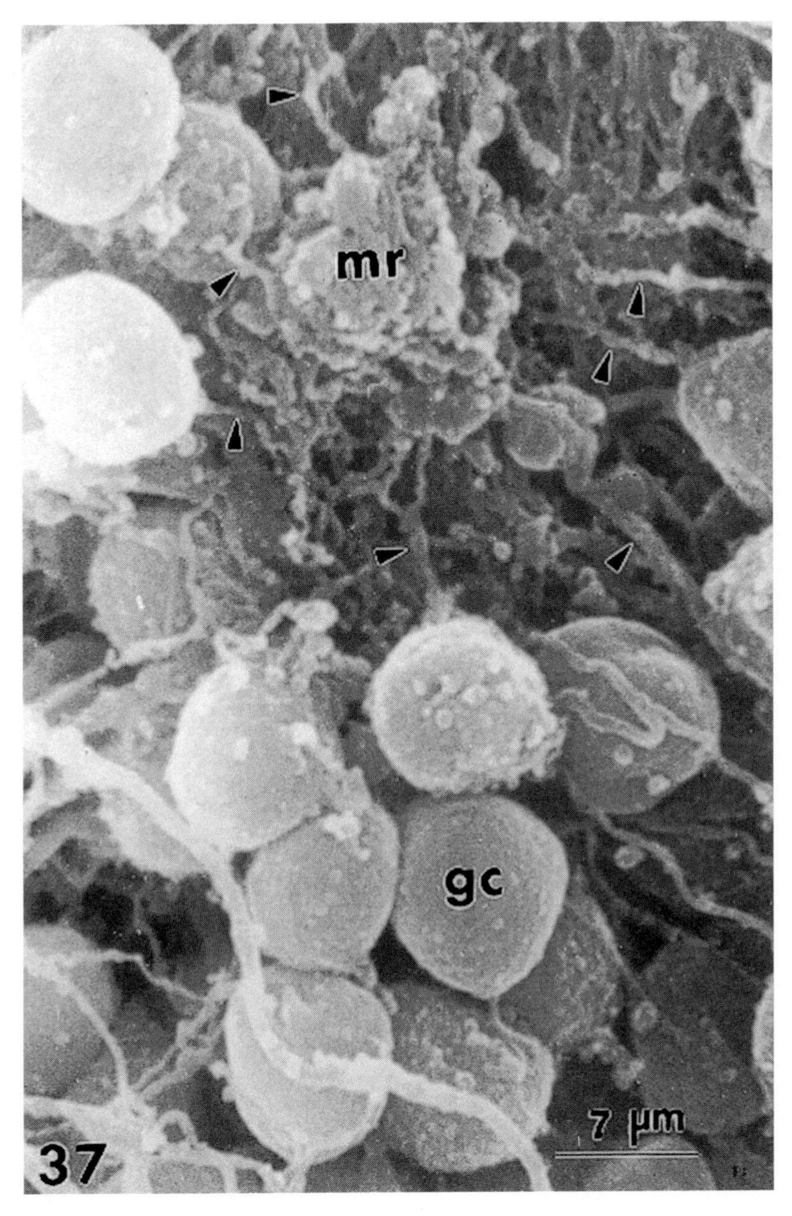

FIGURE 37. Teleost fish cerebellar cortex. Cross-fractured mossy fiber glomerulus showing the central mossy fiber rosette (mr) surrounded with the radially converging dendrites (arrowheads) upon the rosette of as many as 16 neighboring granule cells. Note the inner radial configuration of mossy fiber glomerulus. Scanning electron microscopy (SEM) freeze–fracture method. Gold–palladium coating.

and axonal field of these multisynaptic complexes are displayed with better definition and sharpness than the images obtained with the Golgi light microscopy (LM) technique (Fox et al., 1967a; Palay & Chan-Palay, 1974). These scanning electron micrographs exhibit a new view of cerebellar cortex synaptic architecture which opens promising lines of research for qualitative and quantitative studies of the glomerular region. Recently, a more accurate quantitative relationship between mossy fibers and granule cell dendrites has been shown with computer-assisted microscopy (Hámori, Jakab, & Takacs, 1997) and confocal laser scanning microscopic studies (Castejón et al., 2000a) that reported 50–60 granule cells surrounding the mossy glomerular region.

Chapter 5

Golgi Cells

SHORT HISTORY

The Golgi cells or large stellate cells of the cerebellar granule cell layer have been extensively studied with the light microscope by Golgi (1883), Kölliker (1890), Van Gehuchten (1891), Retzius (1892b), and Ramón y Cajal (1955). Fox (1962), Eccles, Ito, and Szentágothai (1967), Mugnaini (1972), Castejón and Castejón (1972), Palay and Chan-Palay (1974), Castejón (1976), Sturrock (1990), and Alvarez Otero and Anadon (1992), among others, have carried out further electron and light microscopic studies. Some conventional scanning electron microscopy (SEM) observations of Golgi cells have been published in general studies dealing with the cytoarchitectonic arrangement and intracortical circuits of the cerebellar cortex (Castejón, 1981, 1984, 1988, 1993, 1996; Castejón & Apkarian, 1992; Castejón & Caraballo, 1980a, 1980b; Castejón & Castejón, 1991; Castejón & Valero, 1980; Hojo, 1994). More recently, a detailed confocal, light, SEM and transmission electron microscopic (TEM) study of Golgi cells of several vertebrates has been reported (Castejón & Castejón, 2000).

SCANNING ELECTRON MICROSCOPY OF UNFRACTURED GOLGI CELLS

Cryofractured teleost fish cerebellar granular layer examined by conventional SEM, in sample preparations coated with gold–palladium, displays the three-dimensional view of the round Golgi cell soma in the granular layer which appears as a macroneuron, about 25–30 μm in diameter, surrounded by small granule cells 5–10 μm in diameter. Gold–palladium coating, about 5–10 nm thick, produces a rough outer surface on cerebellar nerve cells (Figure 38). Examination at higher magnification of Rhesus monkey cerebellar granular layer with field emission SEM (FESEM) processed with the cryofracture method and coated with chromium, shows the true, smooth cotton-like outer surface of ovoid or round Golgi cell bodies which exhibit an ascending dendrite toward the molecular layer and a horizontal dendrite that remains in the granular layer (Figure 39).

In the granular layer of human cerebellar cortex, prepared with the ethanol-cryofracturing technique and coated with gold–palladium, the image of the outer surface

FIGURE 38. Conventional scanning electron micrograph of teleost fish cerebellar cortex. The Golgi cell soma (go) appears surrounded by granule cells (gc). Note the rough outer surface of both cell types, which is a decorative effect produced by the heavily deposited gold–palladium coating which masks the true outer surface of their plasma membranes.

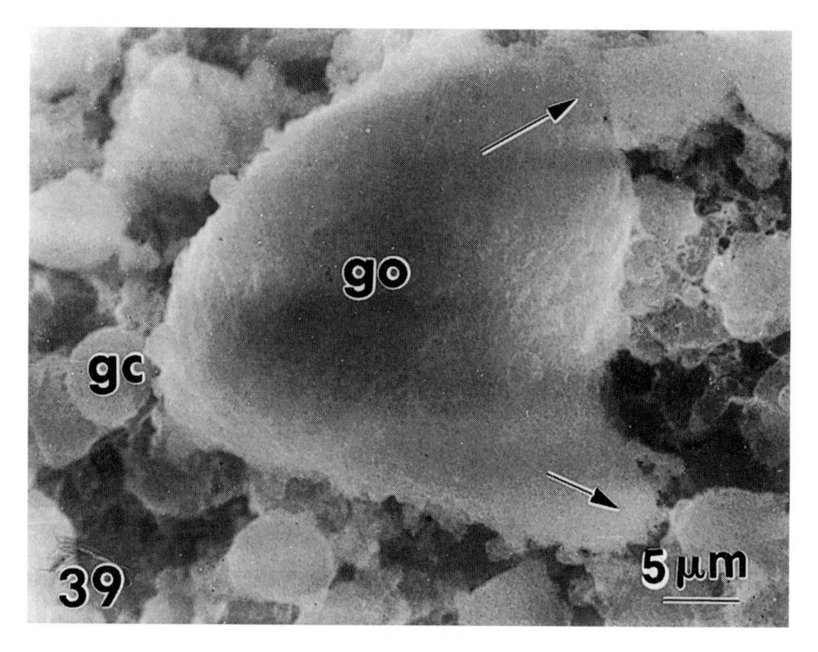

FIGURE 39. Field emission scanning electron micrograph of Rhesus monkey cerebellar cortex. Freeze–fracture scanning electron microscopy (SEM) method. A Golgi cell (go) appears as a large neuron surrounded by numerous granule cells (gc). Note their smooth outer surfaces corresponding to the glycocalyx layer. Chromium coating.

FIGURE 40. Human cerebellar cortex. Ethanol-cryofracturing technique. The cryodissected Golgi cell (go) shows a round body with an ascending dendrite toward the molecular layer (long arrow) and a horizontal dendrite (arrowheads) remaining in the granular layer. The short axonal ramification (short arrows) appears to be insinuating amongst neighboring granule cells (gc). Gold–palladium coating.

of large Golgi cells was obtained by the cryodissecting process induced by this method (Figure 40). This image can be correlated with that obtained with the Golgi light microscopy (LM). The cryofracture process with liquid nitrogen (slow freezing) exposes not only the cell body, but also the ascending and horizontal dendrites and the short axonal ramification. The depth of focus of the SEM permits tracing of the horizontal dendrites and the short axonal ramifications pervading the neighboring neuropil of the granular layer.

SEM examination at high magnification of the Golgi axonal plexus, shows the extension and wavy course of Golgi axonal ramifications in the granular layer (Figure 41).

The beaded shape and fine axonal processes of Golgi cells can be observed in the granular layer, making one to one contact with granule cell dendrites (Figure 42). As formerly described by Ramón y Cajal (1955) and supported later by Hámori and Szentágothai (1966a), some axonal ramifications were traced by SEM and observed to be participating in the formation of mossy glomerular peripheral region (Castejón, 1988; Castejón & Castejón, 2000).

SCANNING ELECTRON MICROSCOPY OF FRACTURED GOLGI CELLS

The freeze–fracture method for SEM using liquid nitrogen (slow freezing) and applied to the study of mouse cerebellum reveals at low magnification the fractured Golgi

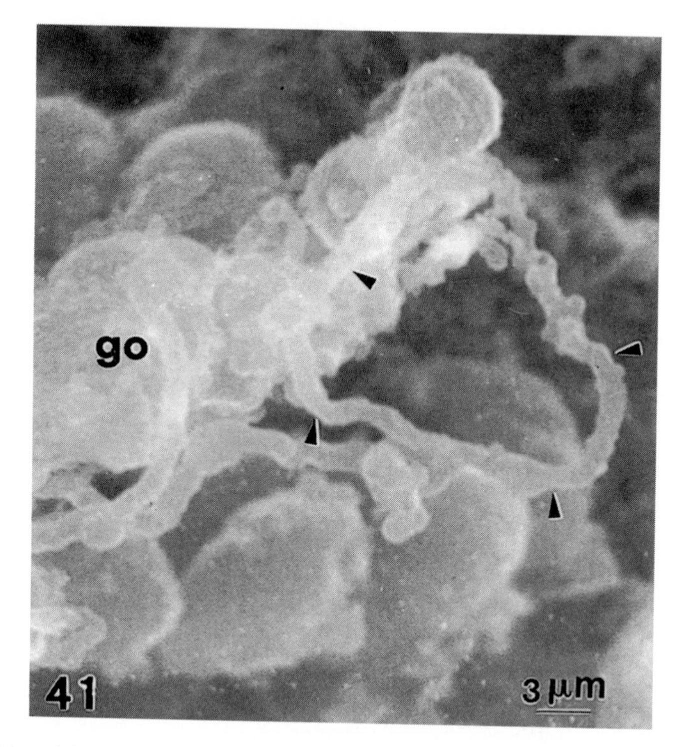

FIGURE 41. Teleost fish cerebellar cortex. Golgi cell (go) showing the stereospatial distribution of its short axonal ramifications (arrowheads) in the granule cell layer. Gold–palladium coating.

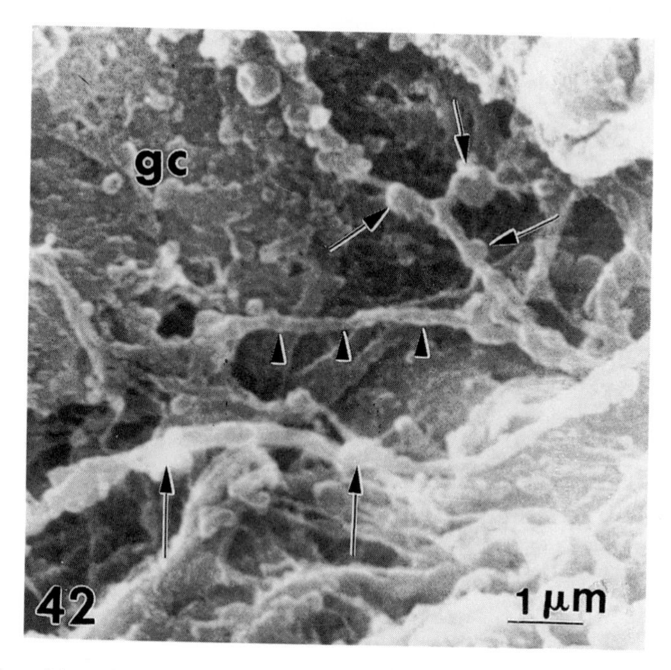

FIGURE 42. Teleost fish cerebellar cortex. Golgi cell beaded shape terminal axonal ramification (short arrows) in the granular layer making one-to-one synaptic relationship with a granule cell (gc) dendrite (arrowheads). The long arrows indicate the longitudinal profile of a neighboring mossy fiber. Gold–palladium coating.

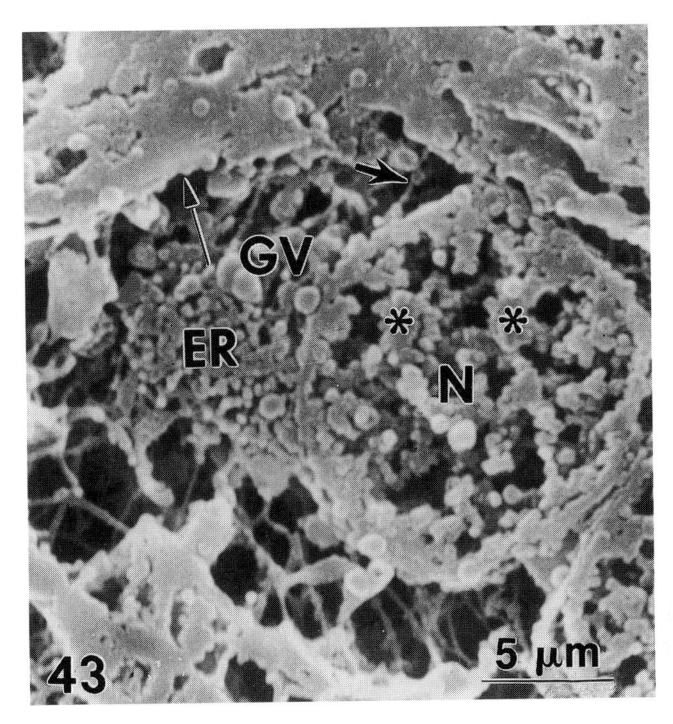

FIGURE 43. Teleost fish cerebellar cortex. Freeze–fracture scanning electron microscopy (SEM) method. Fractured Golgi cell showing the heterochromatin masses (asterisks) of the nucleus (N), the spatial arrangement of endoplasmic reticulum (ER), Golgi complex vesicles (GV), and cytoskeletal structures (short arrow). The long arrow indicates the outline of Golgi cell plasma membrane. Gold–palladium coating.

and granule cells and the three-dimensional arrangement and inner details of fractured cytoplasm and nucleus (Figure 43). The outer surface and stereospatial arrangement of the Golgi cell rough endoplasmic reticulum, cellular organelles such as the Golgi complex, and also of the nuclear chromatin substance are seen. In teleost fish samples, the cryofracture method and the "washing out" process of the soluble fraction of the cytoplasm or cytosol made possible, during the thawing step, the visualization of the outer surface of cytomembranes. The SEM images can be easily compared to TEM micrographs shown by Mugnaini (1972), Palay and Chan-Palay (1974), and Castejón and Castejón (2000).

CONCLUDING REMARKS

The three-dimensional morphology and outer surface of Golgi cells can be disclosed with the ethanol-cryofracturing technique. This method shows their horizontal dendrites and short axonal ramifications in the granular layer and the ascending dendrites directed toward the molecular layer. The depth of focus of SEM allows tracing of the beaded axonal ramifications in the granular layer and their synaptic relationships with granule cell dendrites at the periphery of glomerular region. The freeze–fracture method for SEM exposed the Golgi cell cytoplasmic and nuclear compartment organization.

Chapter 6

Unipolar Brush Cells

RECENT HISTORY AND THREE-DIMENSIONAL MORPHOLOGY

Muñoz (1990) described monodendritic neurons in the human cerebellar cortex using chromogranin A-like immunoreactivity. Mugnaini and Floris (1994), Mugnaini, Floris, and Wright-Gross (1994), Mugnaini, Diño, and Jaarsma (1997) with antibodies to calretinin and neurofilament proteins characterized similar cells which they termed unipolar brush cells (UBC) in the granular layer of mammalian cerebellar cortex. These cells form conspicuous and elaborated asymmetric synapses with one or two mossy fiber rosettes. The cell body displays a deeply indented nucleus, large Golgi apparatus, numerous filaments, microtubules, mitochondria and large, dense core vesicles. The cell body gives rise to a thin axon and a single large dendritic trunk, the tip of which forms a tightly packed group of branches that resemble a paintbrush as they contact the mossy rosettes. They also form symmetric and asymmetric synaptic junctions with presumed Golgi cell boutons. Acetylcholine and glutamate are apparently co-released at these synapses. At the mossy glomerulus, the UBC dendrites are intermingled with the granule cell dendrites. Their dendrioles form dendro–dendritic contacts with granule cell dendrites (Mugnaini et al., 1997).

The UBC in teleost fish cerebellar cortex appear as round or oval cells closely resembling the morphology of granule cells. However, they can be characterized by their unique dendritic process that ends in a triangular or conical expansion (Figure 29). This process appears topographically related to the mossy fiber and their rosette formations.

FUTURE RESEARCH

Three-dimensional morphology of unipolar brush cells should be systematically examined using confocal laser scanning microscopy and scanning electron microscopy (SEM) employing the cryofracture technique, at the level of the cerebellar granular layer.

Chapter 7

Lugaro Cells

SHORT HISTORY

Fox (1959) described Lugaro cells as spindle-shaped cells transversely oriented in the granular layer and located immediately beneath the Purkinje cell layer. He observed their horizontally directed dendrites in connection with the basket cell descending axonal collaterals, forming the *pinceaux* around the axon hillock region of the Purkinje cell. Fox also traced descending dendrites in synaptic relationship with mossy fiber rosettes, at the level of mossy glomerulus, and with the Golgi cell axonal ramifications. According to Fox, the axonal collaterals of Lugaro cells contact the basket cell bodies in the lower molecular layer. Palay and Chan-Palay (1974) reported two types of horizontal fusiform cells possessing different axonal pattern and distribution. One type gives rise to an axon directed downward to the granular layer and reaching the white matter as described by Ramón y Cajal (1955). Sahin and Hockfield (1990) used double label immunocytochemistry to identify and classify the Lugaro cells in cat cerebellum. Laine and Axelrad (1996, 1998) described Lugaro cells with axons projecting to the molecular and granular layers and targeting stellate and basket cells. Melik-Musyan and Fanardzhyan (1998) demonstrated two Lugaro cell types with fusiform and triangular cell bodies. According to these authors, Lugaro cells also exhibit projection fibers traveling from the molecular layer to the white substance. Using electronmicroscopic gold-toning procedure and post-embedding anti-GABA immunocytochemistry, they demonstrated that Lugaro cell axon forms multiple symmetrical synaptic connections with basket and stellate cell soma and proximal dendrites. In addition, partially myelinated Lugaro cell axons form a parasagittal plexus and extend long transverse branches that run longwise to the parallel fibers. These transverse axons participate in the synaptic contacts of Lugaro cells with Golgi cells. Negyessi, Vidnyanzky, Kuhn, Knöpfel, Gores, and Hamori, (1997) used light microscopy (LM) and transmission electron microscopy (TEM) pre-embedding immunoperoxidase and immunogold methods to demonstrate the cellular and subcellular localization of mGluR5 metabotropic glutamate receptor in Golgi and Lugaro cells as well as in parallel fiber synaptic contacts. Okhotin and Kalinichenko (1999) reported dendro–somatic and somato–somatic contacts of Lugaro cells and dendro–dendritic contacts between Lugaro and Golgi cells.

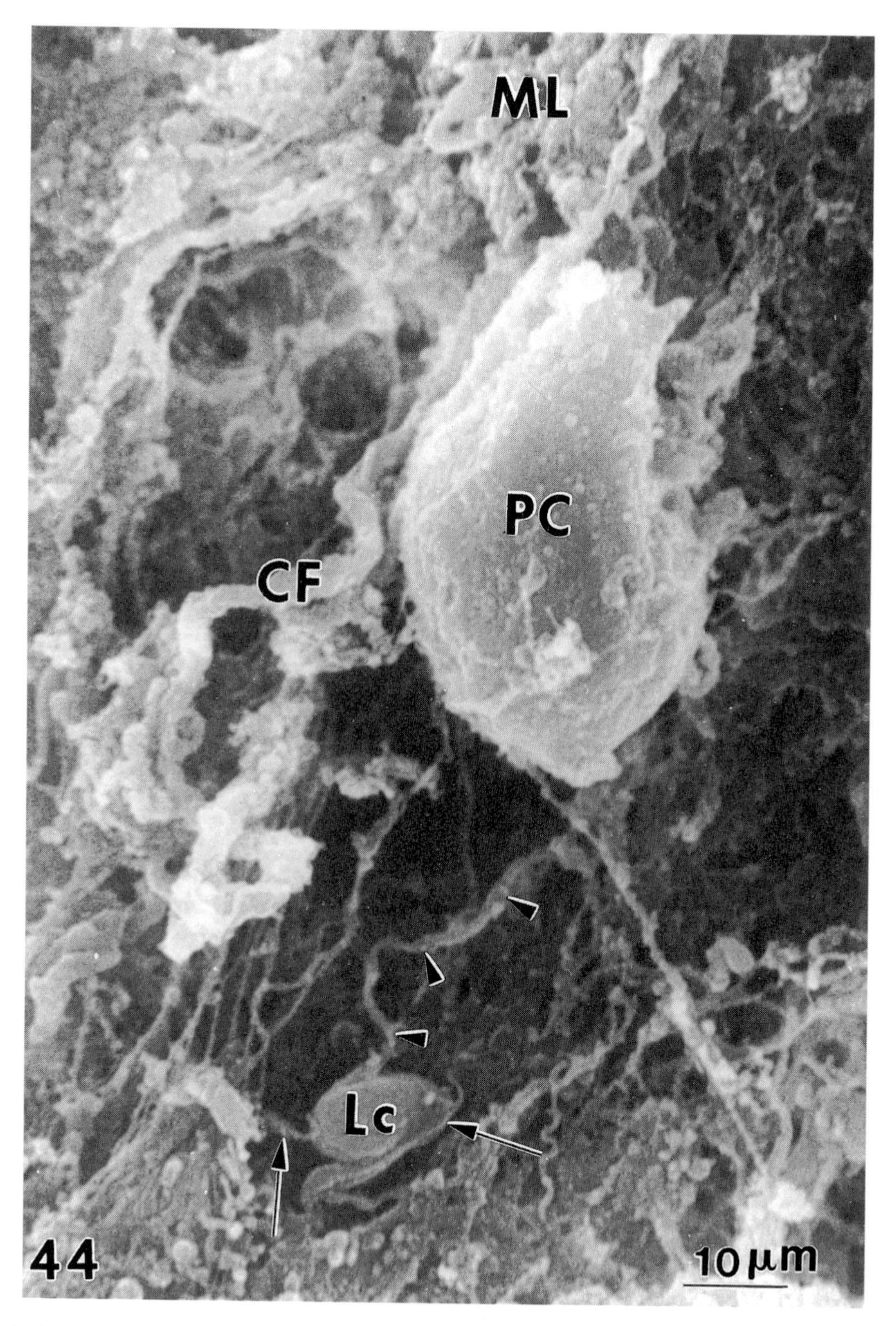

FIGURE 44. Cryofractured teleost fish cerebellar cortex. Freeze–fracture scanning electron microscopy (SEM) method. Lugaro cell (Lc) showing an ovoid cell body and an ascending axon (arrowheads) directed toward the molecular layer and horizontal dendrites (arrows) remaining in the granular layer. A climbing fiber (CF) is also observed approaching a Purkinje cell soma (PC) as it ascends to the molecular layer (ML). Gold–palladium coating.

THREE-DIMENSIONAL MORPHOLOGY

In teleost fish cerebellar cortex examined by conventional scanning electron microscopy (SEM), Lugaro cells appear as ovoid cells located underneath Purkinje cells. They show an ascending axon directed toward the molecular layer and horizontal dendrites that remain in the granular layer (Figure 44).

FUTURE RESEARCH

Lugaro cells should be systematically examined using modern three-dimensional microscopy techniques such as confocal laser scanning microscopy and field emission SEM (FESEM) using cryofracture technique.

Chapter 8

Purkinje Cells

THE LONG HISTORY

Purkinje cells were first described using light microscopy (LM) by Purkinje (1837) and subsequently studied by Denissenko (1877), Golgi (1882, 1883, 1885), Fusari (1883, 1887), Obersteiner (1888), Ramón y Cajal (1890a), Kölliker (1890), Retzius (1892a, 1892b), Falcone (1893), Dogiel (1896), Popoff (1896), Smirnov (1897), Held (1897), Crevatin (1898), Bielschowsky and Wolff (1904), Lache (1906), Estable (1923), Jakob (1928), Fox and Barnard (1957), Fox (1962), and Fox, Siegesmund, and Dutta (1964). With the advent of transmission electron microscopy (TEM) mainly Palay and Palade (1955), Gray (1961), Herndon (1963), Hámori and Szentágothai (1964, 1968), Castejón (1968), Hillman (1969), Mugnaini (1972), Palay and Chan-Palay (1974), and Hillman and Chen (1981) meticulously examined the Purkinje cells in several vertebrates. More recently, a detailed microscopic analysis of the structural and synaptic organization of Purkinje cells have been performed by numerous investigators (Chen & Hillman, 1999; Harvey & Napper, 1988; Hirokawa, 1989; Ji & Hawkes, 1994; Kanaseki, Ykeuchi, & Tashiro, 1998; Khan, 1993; Kosaka, Kosaka, Nakayama, Hunziker, & Heiszmann, 1993; Martone, Zhang, Simpliciano, Carrager, & Ellisman, 1993; Matsumura & Kohno, 1991; Meek & Nieuwenhuys, 1991; Meller, 1987; Monteiro, Rocha, & Marini-Abreu, 1994; Rusakov, Podini, Villa, & Meldolesi, 1993; Tandler, Rios, & Pellegrino de Iraldi, 1997; Terasaki, Slater, Feiw, Schmidek, & Reese, 1994; Teune, Van Der Burg, De Zeeuw, Voogd, & Ruigrok, 1998).

Conventional scanning electron microscopy (SEM) and high resolution field emission SEM (FESEM) have provided a new view of the Purkinje cell outer surface, its three-dimensional morphology, and synaptic connections (Castejón, 1988, 1990a, 1990c, 1991, 1996; Castejón & Apkarian, 1992, 1993; Castejón & Caraballo, 1980a, 1980b; Castejón & Castejón, 1988; Castejón & Valero, 1980; Castejón, Castejón, & Apkarian, 1994a, 1994c; Castejón, Apkarian, & Valero, 1994b; Reese, Landis, & Reese, 1985). Using a creative tearing technique, Scheibel, Paul, and Fried (1981), exposed the outer surface of Purkinje cells, the surrounding basket cell axon collaterals and segments of climbing fibers. By means of ultrasonic microdissection, Arnett and Low (1985) showed Purkinje cells, basket cell synapses, and Purkinje dendritic spines. Takahashi-Iwanaga (1992) showed the reticular endings of Purkinje cell axons in the rat cerebellar nuclei, by means of NaOH maceration. Hojo (1994) prepared specimens of human cerebellar cortex, by means of t-butyl alcohol freeze-drying device to examine the Purkinje cell somatic surface. Castejón and Castejón (1997) described, in detail, the SEM three-dimensional features of Purkinje cells and their synaptic relationship in the molecular layer.

THREE-DIMENSIONAL MORPHOLOGY AND OUTER SURFACE

Exploration of the Purkinje cell layer of teleost fishes, at higher magnifications by means of conventional SEM, and using a 20 nm high resolution scanning probe, shows the outer surface of Purkinje cells (Figures 45 and 46). The Bergmann cell cytoplasm often is removed during the slicing procedure for conventional SEM thereby exposing the Purkinje cell outer surface. The Purkinje cell outer surface exhibits a rough appearance due to attached vestiges of Bergmann cell cytoplasm and axosomatic synaptic endings of the neighboring basket cells which form the Purkinje pericellular nest. The Purkinje cell soma shows a typical pyriform, flask, or round shape. By means of the slicing technique in the teleost fish cerebellum, and concerning Purkinje cells, it was possible to obtain a view of

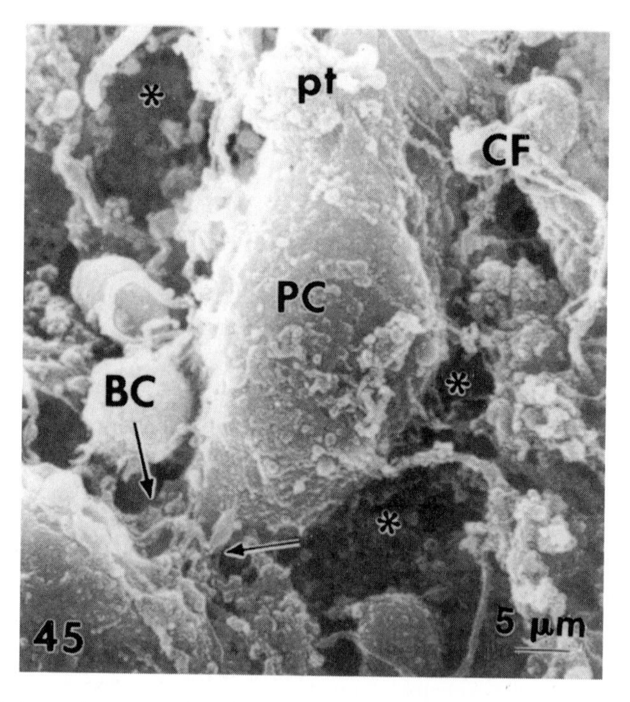

FIGURE 45. Sagittaly cryofractured teleost fish cerebellar cortex, showing the outer surface of a Purkinje cell (PC) soma. The cryofracture process has removed the ensheathing Bergmann cell cytoplasm, exposing the outer surface of the pear-shaped neuronal soma. Note the origin of the primary dendritic trunk (pt). Climbing fibers (CF) are seen approaching the primary dendritic trunk. The asterisks label the dark spaces previously occupied by Bergmann cell cytoplasm. A basket cell (BC) sends its axon toward the Purkinje cell soma. The arrows indicate the Purkinje cell infraganglionic plexus. Gold–palladium coating.

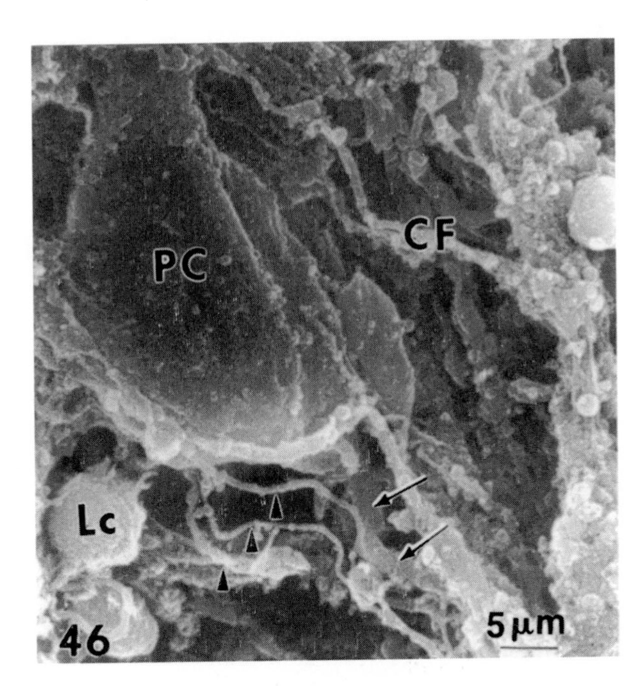

FIGURE 46. Sagittaly cryofractured teleost fish cerebellar cortex showing the elongated Purkinje cell body (PC), its axon hillock region, and axonal initial segment (arrows). The Purkinje cell infraganglionic plexus (arrowheads) is also seen. A small neuron, presumably a Lugaro cell (Lc), is observed at the lower left corner of the figure. A climbing fiber (CF) is also distinguished. Gold–palladium coating.

the lower pole, the axon hillock region or axon attachment zone and the infraganglionic plexus of Purkinje cells. Due to the section and cryofracture procedures used in the present study, it was very difficult to trace the descending course of Purkinje cell axons through the granular layer. Some partial views were obtained at the level of white matter as described in Chapter 2. The Purkinje cell axonal recurrent collaterals and their synapses, as described by Hámori and Szentágothai (1968) using TEM, is an important issue that should be investigated at the SEM level. The emergence and outer surface of the primary dendritic trunk was observed at the upper pole of the cell, ascending toward the molecular layer. The immediate fixation in situ by vascular perfusion with 4% glutaraldehyde procedure preserved fairly well the three-dimensional relief of the Purkinje cell body avoiding, to a certain extent, the collapse or shrinking process. With FESEM, using delicate handling of the tissue and chromium coating, Bergmann cell cytoplasm is well preserved like a veil of high topographic contrast encapsulating the Purkinje cell soma (Figure 47).

The SEM freeze–fracture method applied to teleost fishes was useful in obtaining longitudinal (Figures 48 and 49) and cross fractures (Figure 50) of Purkinje secondary dendritic ramifications. The endoplasmic reticulum profiles and cytoskeletal elements form a microtrabecular arrangement in which mitochondria appear suspended (Castejón & Castejón, 1997). This microtrabecular arrangement is also observed in cryofixed Purkinje

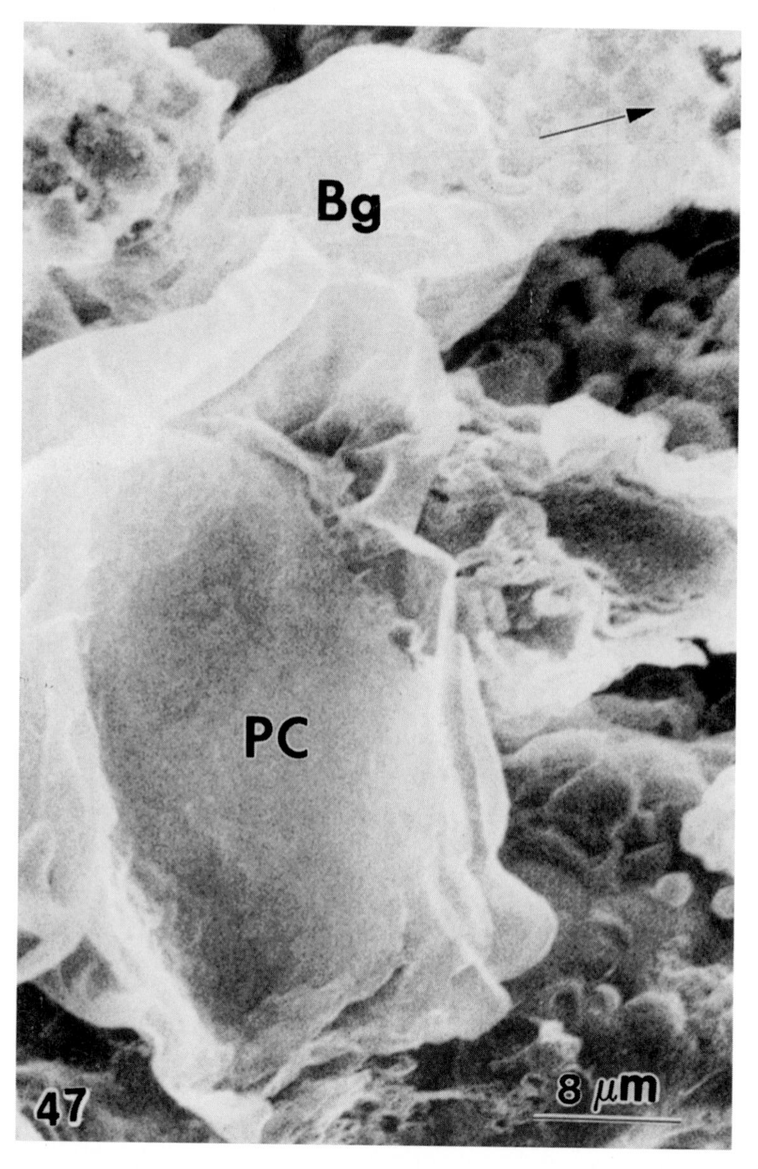

FIGURE 47. Field emission high resolution scanning electron micrograph of cryofractured mouse cerebellar cortex. A thin sheath of the high density Bergmann cell cytoplasm (Bg) envelopes the Purkinje cell soma (PC). The ascending Bergmann fiber (arrow) also is observed ascending to the molecular layer. Chromium coating.

cells of chicken cerebellum (Meller, 1987). The cytoplasmic soluble fraction is "washed out" during the thawing step of the SEM cryofracture process, thus allowing the visualization of cytomembranes and cell organelles. Gold–palladium coating clearly and heavily delineates the longitudinal and cross-sections of the non-synaptic segments of parallel fibers in the molecular layer.

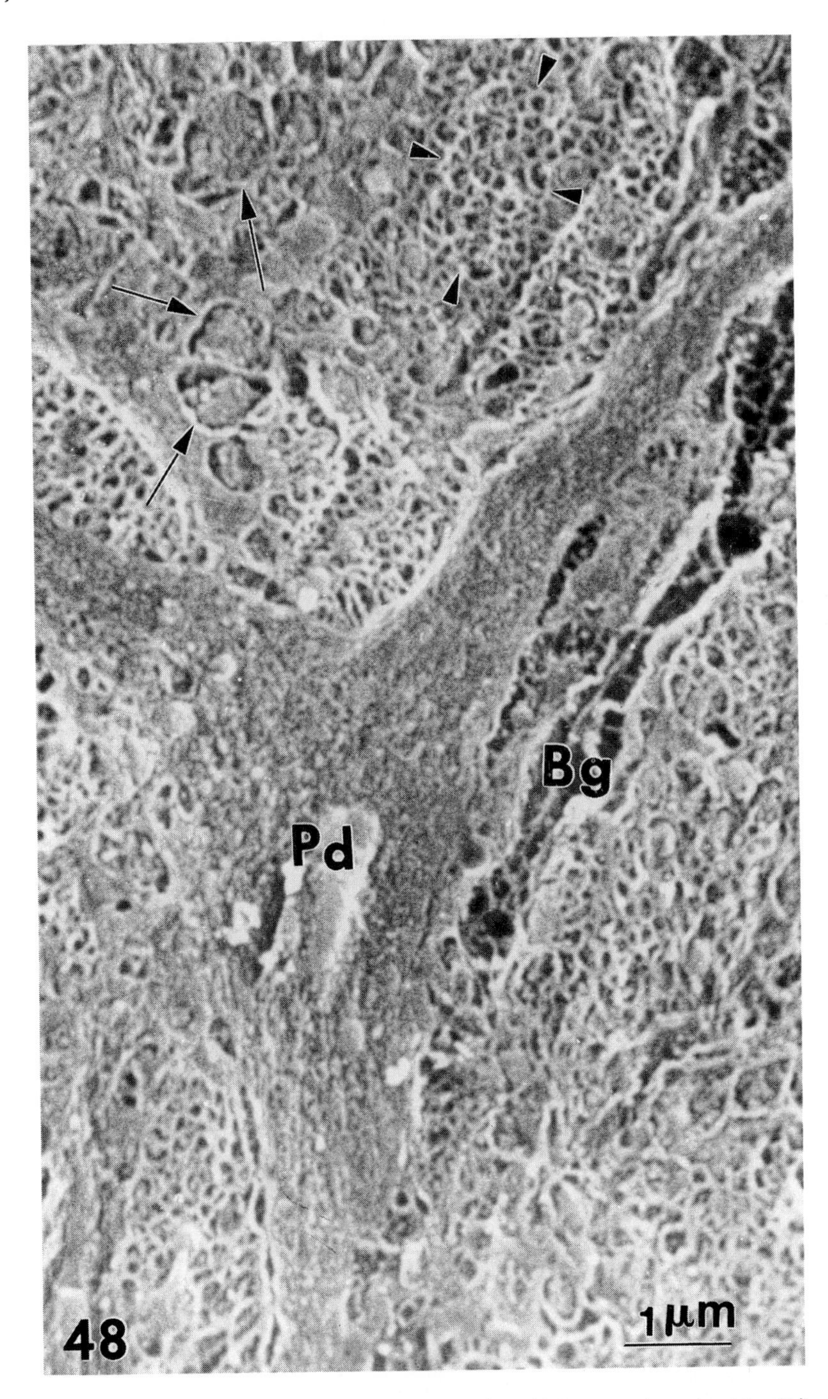

FIGURE 48. Teleost fish cerebellar cortex. Cryofractured Purkinje secondary dendrite (Pd) showing the associated Bergmann cell cytoplasm (Bg). The non-synaptic segments (arrowheads) and synaptic varicosities (arrows) of parallel fibers containing spheroid synaptic vesicles are seen in the neighboring neuropil. Gold–palladium coating.

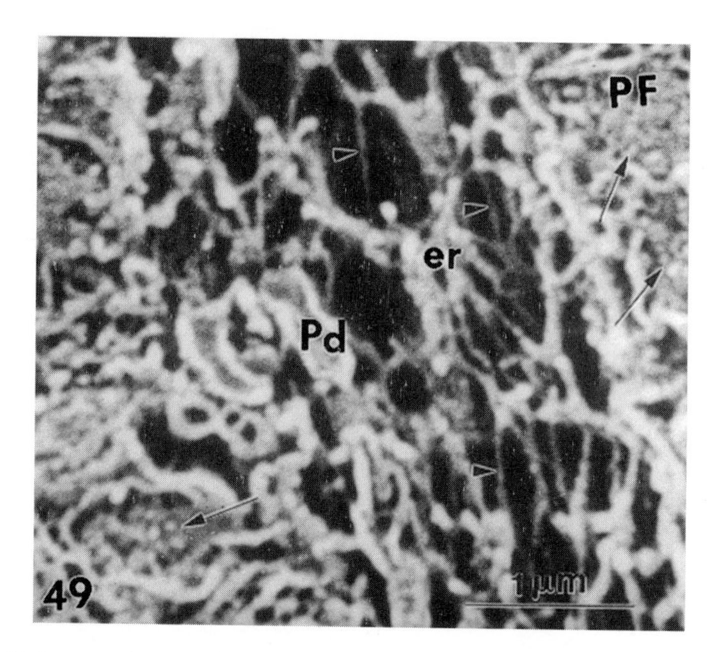

FIGURE 49. Higher magnification of teleost fish Purkinje cell molecular layer. Freeze–fracture scanning electron microscopy (SEM) method. Longitudinally fractured secondary dendrite (Pd) showing the trabecular system formed by the endoplasmic reticulum (er) and cytoskeletal structures (arrowheads). The neighboring fractured parallel fiber synaptic varicosities (PF) exhibit the spheroid synaptic vesicles (arrows). Gold–palladium coating.

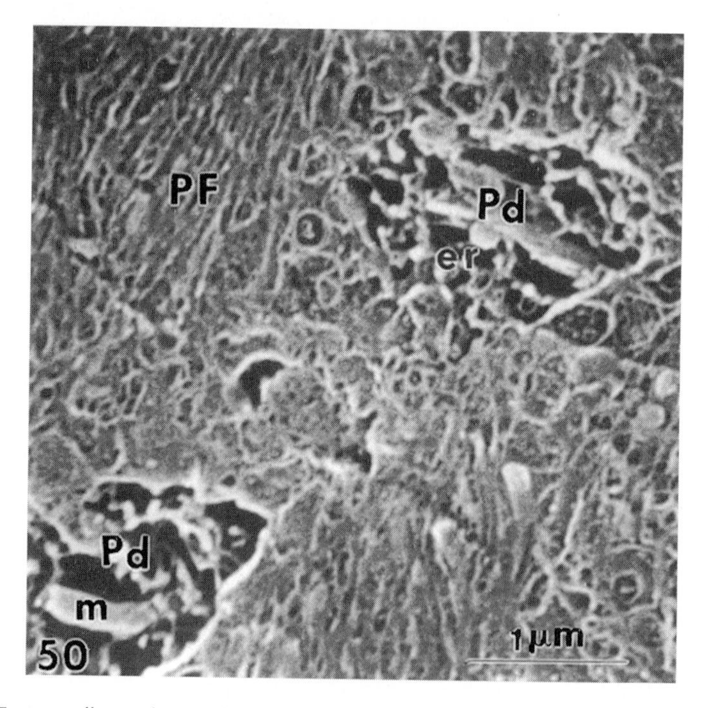

FIGURE 50. Transversally cryofractured teleost fish cerebellar molecular layer showing the cross-sections of two Purkinje secondary dendritic branches (Pd), surrounded by the longitudinal profiles of parallel fiber bundles (PF). Note the elongated mitochondrion (m) and the trabecular system formed by endoplasmic reticulum profiles (er). Freeze–fracture scanning electron microscopy (SEM) method. Gold–palladium coating.

SYNAPTIC RELATIONSHIP WITH PARALLEL FIBERS

The parallel fibers or granule cells axons also were traced in the human cerebellar molecular layer processed by the ethanol-cryofracturing technique. They show *en passant*, axo-dendritic synapses with many successive Purkinje dendrites (Figure 51). The non-synaptic segments of parallel fiber appear as fine, high mass density threads applied tangentially to the outer surface of Purkinje dendrites. The synaptic segments of parallel fibers are closely applied to the bulbous soma of Purkinje dendritic spines making crossing over, *en passant* axo-spinodendritic contacts. This view could be easily correlated with stereodiagrams made by Hámori and Szentágothai (1964), and Ito (1984) from earlier thin section EM studies. The SEM three-dimensional morphology supports previous descriptions on the geometrical arrangement and topographic relationship between pre- and post-synaptic structures of parallel fiber–Purkinje spine synapses made by TEM (Castejón, 1968, 1981, 1988, 1990a; Gray, 1961; Landis & Reese, 1974; Mugnaini, 1972; Palay & Chan-Palay, 1974).

The longitudinal profiles of secondary and tertiary spiny Purkinje dendrites (Figure 52) in the cerebellar molecular layer can be studied by FESEM. Unattached Purkinje dendritic spines can be examined showing the spine body and neck (Figure 53). Close examination of the cerebellar molecular layer discloses the three-dimensional morphology and outer

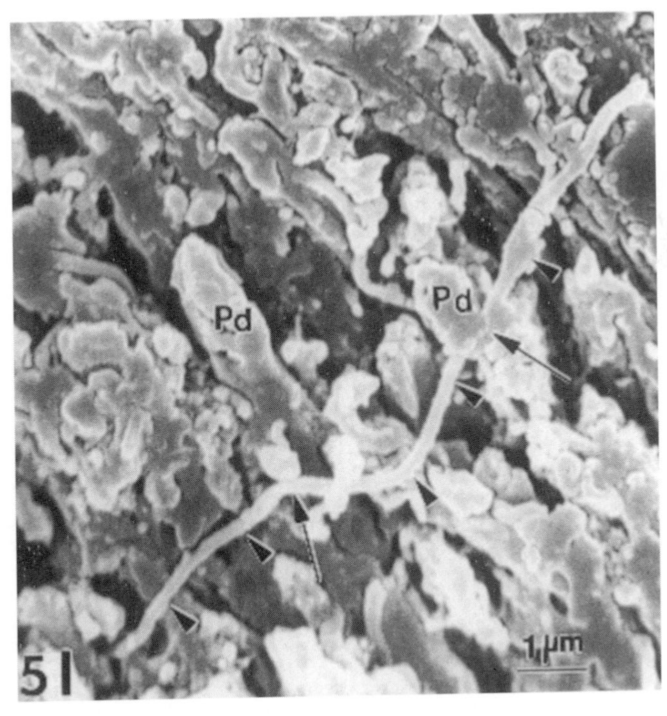

FIGURE 51. Human cerebellar cortex processed by the ethanol-cryofracturing technique. A parallel fiber (arrowheads) is making *en passant* synaptic contacts (arrows) with successive Purkinje dendrites (Pd). Gold–palladium coating.

FIGURE 52. Low magnification field emission scanning electron microscopy of a cryofractured mouse cerebellar cortex. The longitudinal profiles of tertiary Purkinje dendritic branches (arrows) and their spines (arrowheads) are observed. Remnants of Bergmann cell cytoplasm (Bg) are also noted. Chromium coating.

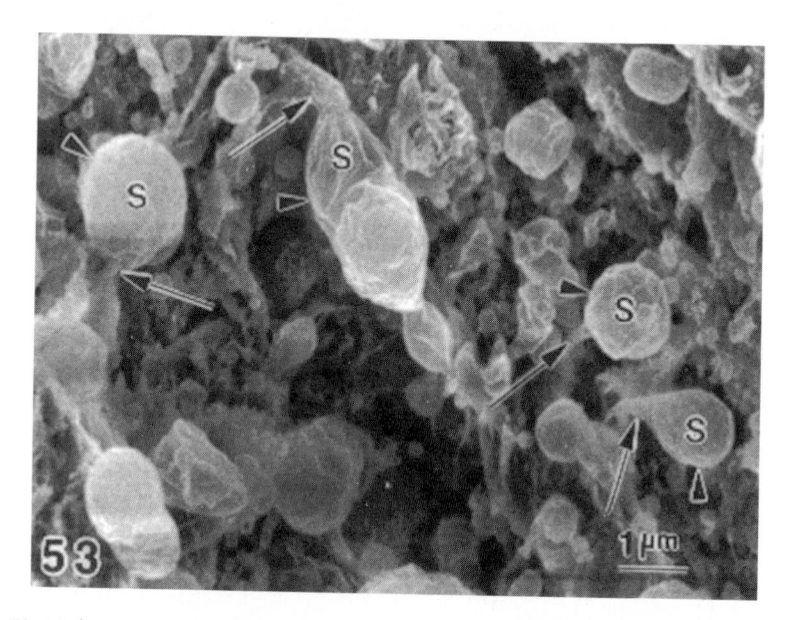

FIGURE 53. High magnification field emission scanning electron microscopy of mouse cerebellar molecular layer showing unattached Purkinje dendritic spines (S). The cryofracture process removed the neighboring parallel fibers and exposed the spine bodies (arrowheads) and necks (arrows). Chromium coating.

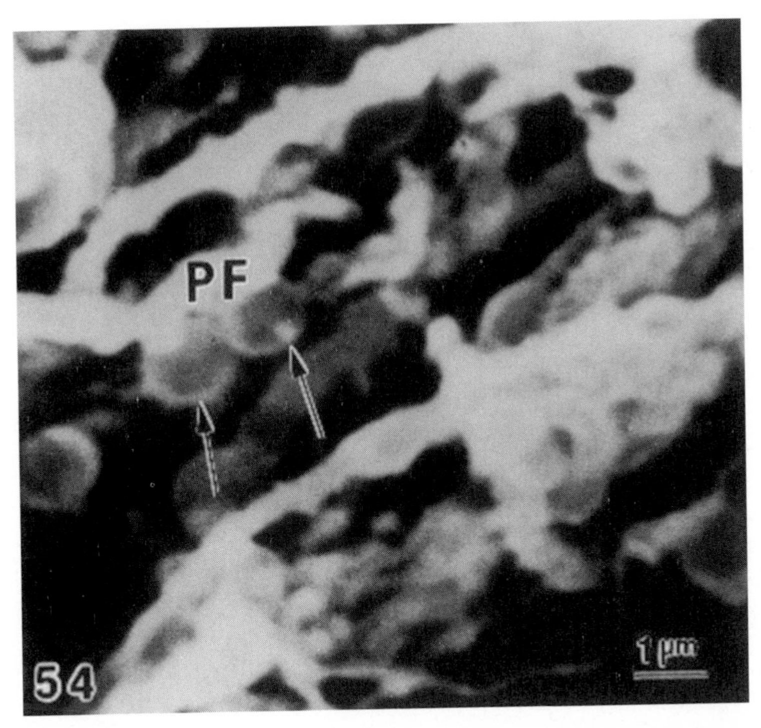

FIGURE 54. Conventional scanning electron microscopy micrograph of teleost fish cerebellar molecular layer showing the outer surface of a parallel fiber (PF) making successive and *en passant* synaptic contacts with two Purkinje dendritic spine bodies (arrows). SEM slicing technique. Gold–palladium coating.

surface of parallel fiber–Purkinje dendritic spine synapses (Figure 54). The FESEM and the freeze–fracture method for SEM show at intermediate magnifications, the fractured parallel fiber synaptic varicosities containing clustered spheroid synaptic vesicles (Figure 55). In addition, the three-dimensional relief and type I secondary electron (SE-I) image of synaptic vesicles and their clustered arrangement can be appreciated.

The FESEM helped us to elucidate parallel fiber–Purkinje spine synaptic relationship (Castejón, Castejón, & Apkarian, 2001c). SE-I images can be obtained by utilizing delicate specimen preparation procedure, the cryofracture method in Freon 22 cooled by liquid nitrogen, the use of chromium coating, and the "in lens" system of FESEM SE-I images. Fractured synaptic varicosities of parallel fibers show the spheroid synaptic vesicles and their limiting plasma membrane. Fractured Purkinje spines display the profiles of endoplasmic reticulum.

In an attempt to resolve its structure and three-dimensional organization (Castejón & Castejón, 1997), the parallel fiber–Purkinje spine synaptic relationship was studied at high magnification in Rhesus monkey cerebellar cortex. The parallel fiber presynaptic ending (Figure 56) exhibits the topographic contrast of clustered spheroid synaptic vesicles, about 50 nm in diameter, which appear surrounded by a homogeneous extravesicular material. Analysis of the synaptic membrane complex showed the SE-I profile of the pre- and postsynaptic membranes and the synaptic cleft. The postsynaptic membrane and the associated

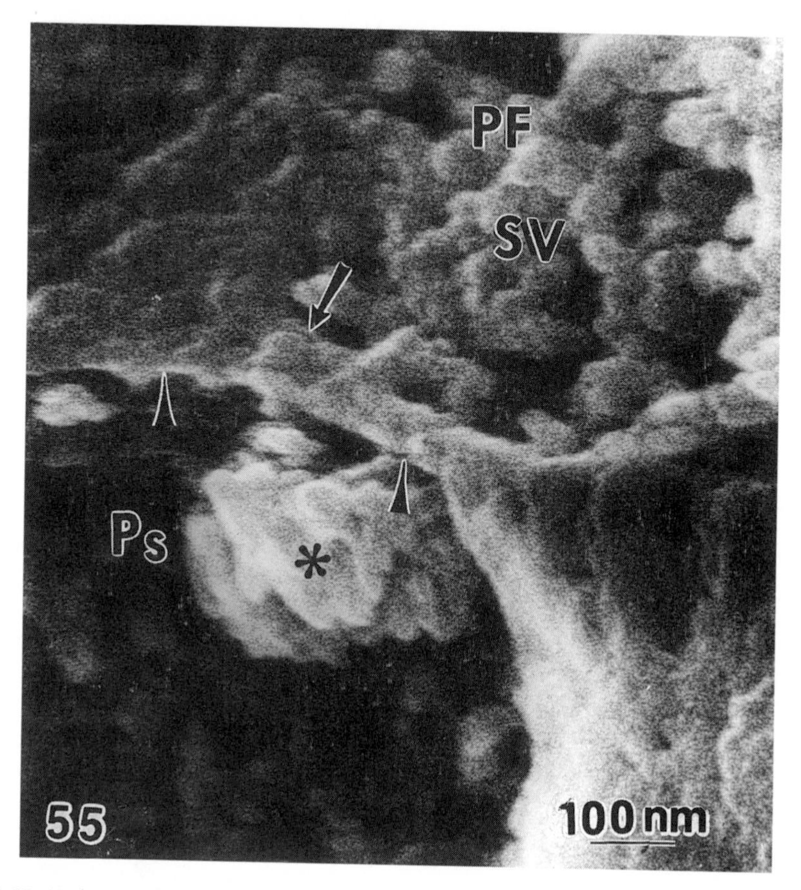

FIGURE 55. High magnification field emission scanning electron microscopy micrograph of Rhesus monkey cerebellar cortex showing a parallel fiber (PF)–Purkinje spine (Ps) synaptic contact. The arrowheads indicate the specialized synaptic membrane complex. Clustered spheroid synaptic vesicles (SV) are seen in the parallel fiber presynaptic ending (PF). Some synaptic vesicles appear anchored to the presynaptic membrane (arrow). The asterisk labels a partial view of Purkinje spine postsynaptic density. Chromium coating.

postsynaptic density showed a discontinuous surface formed by round subunits 25–35 nm in diameter. The SE-I image mode, obtained with the high resolution FESEM, is particularly valuable for providing surface views of the fractured synaptic membrane complex and the synaptic cleft. The SE-I images of postsynaptic subunits correspond, by their topographic localization, to the granular material observed at the postsynaptic density in TEM ultrathin sections. These subunits could also be correlated with the aggregated intramembrane particles (IMPs) localized in the e-face postsynaptic membrane in freeze-etching preparations (Castejón, 1988, 1990a). Their localizations correspond to postsynaptic proteins and/or neurotransmitter receptors (Castejón & Apkarian, 1992).

The synaptic relationship of Purkinje cells with climbing fibers, basket and stellate cells will be described in the following chapters.

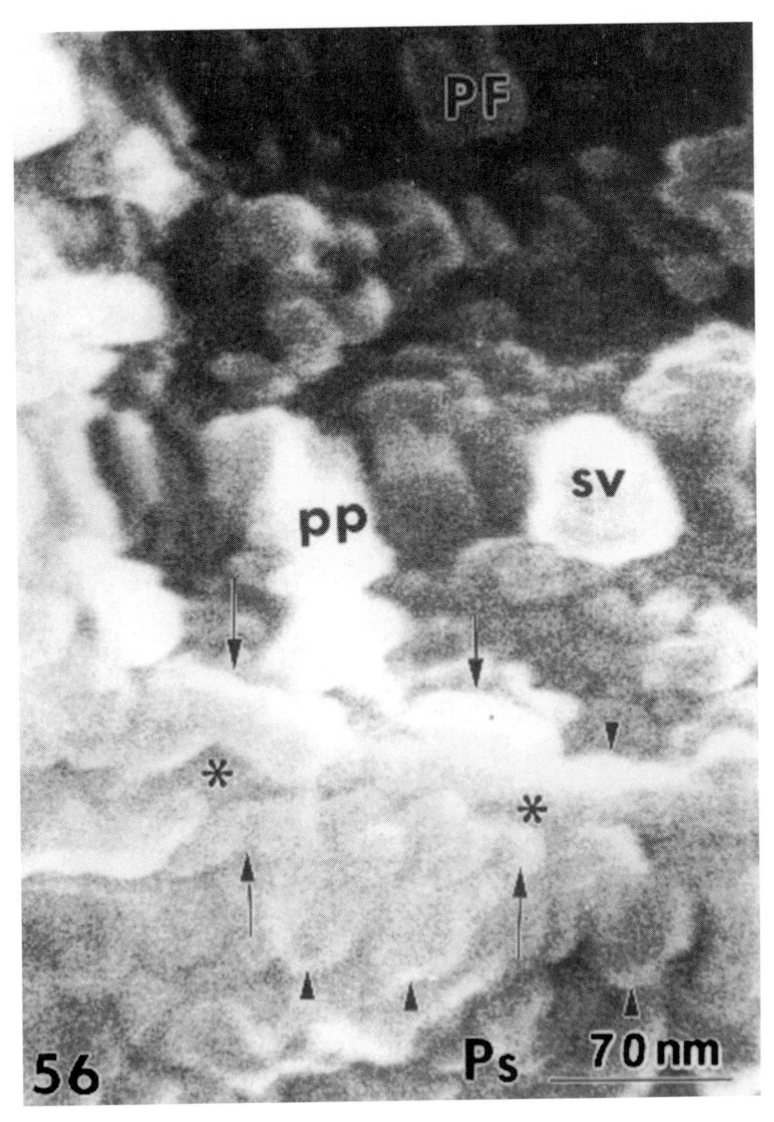

FIGURE 56. High magnification field emission scanning electron microscopy micrograph of parallel fiber (PF)–Purkinje spine (Ps) synaptic membrane complex. The synaptic ending of the parallel fiber shows a presynaptic dense projection (pp) and a synaptic vesicle (sv). The synaptic cleft (asterisks) is seen separating pre- and postsynaptic membranes (arrows). Round globular subunits (arrowheads) are observed associated to the postsynaptic membrane (arrows) corresponding to postsynaptic receptor and/or postsynaptic proteins. Chromium coating.

CONCLUDING REMARKS

Conventional SEM and FESEM exposed the outer surface of Purkinje cells and their synaptic contacts with basket cell axons and parallel fibers. The SEM freeze–fracture method showed the three-dimensional arrangement of endoplasmic reticulum, cytoskeletal structures, and cell organelles. The outer and inner surfaces of crossing over synaptic connections of parallel fibers with Purkinje dendritic spines were also characterized in the molecular layer. High magnification and high resolution FESEM resolved the synaptic membrane complex and also displayed the globular subunits that correspond to the neurotransmitter receptors and/or postsynaptic proteins.

Chapter 9

Climbing Fibers

BRIEF HISTORY

The climbing fibers were first described by Ramón y Cajal (1888) and later by Van Gehuchten (1891), Retzius (1892b), Held (1897), Athias (1897), Estable (1923), and Jakob (1928). Light microscopy (LM) and transmission electron microscopic (TEM) studies of vertebrate cerebellar cortex (Carrea, Reissig, & Mettler, 1947; Chan-Palay & Palay, 1970, 1971a, 1971b; Fox, Andrade, & Schwyn, 1967b; Larramendi & Victor, 1967; Murphy, O'Leary, & Corntlath, 1973; O'Leary, Petty, Smith, & Inukai, 1968; O'Leary, Inukai, & Smith, 1971; Palay & Chan-Palay, 1974; Rivera-Domínguez, Mettler, & Novack, 1974; Scheibel & Scheibel, 1954; Szentágothai & Rajkovits, 1959) have supported and extended the basic morphological aspects of climbing fibers as reported by the pioneering account of Ramón y Cajal (1888, 1955).

More recently, we have described by means of conventional scanning electron microscopy (SEM), and ethanol-cryofracturing technique and freeze–fracture methods, the course of climbing fibers through the cerebellar layers in the mouse, rat, fish, and human cerebellum (Castejón, 1983b, 1986; Castejón & Caraballo, 1980a, 1980b; Castejón & Castejón, 1988; Castejón & Valero, 1980; Castejón, Castejón, & Alvarado, 2000b).

INTRACORTICAL COURSE

Low magnification examination of teleost fish cerebellar white matter with the scanning electron beam in samples coated with gold–palladium, shows the longitudinal bundles of thick afferent mossy parent fibers intermingled with bundles of thin afferent climbing fibers (Figure 57). At higher magnification, both types of afferent fibers could be clearly distinguished by their different thickness. As previously mentioned, the mossy fibers are 2.5 μm in diameter and the climbing fibers measure 1 μm in diameter. Afferent mossy and climbing fibers are additionally distinguished by their branching pattern as they enter the granular layer. The mossy fibers exhibit a characteristic dichotomous pattern of bifurcation whereas climbing fibers display a typical arborescence or crossing-over type of bifurcation.

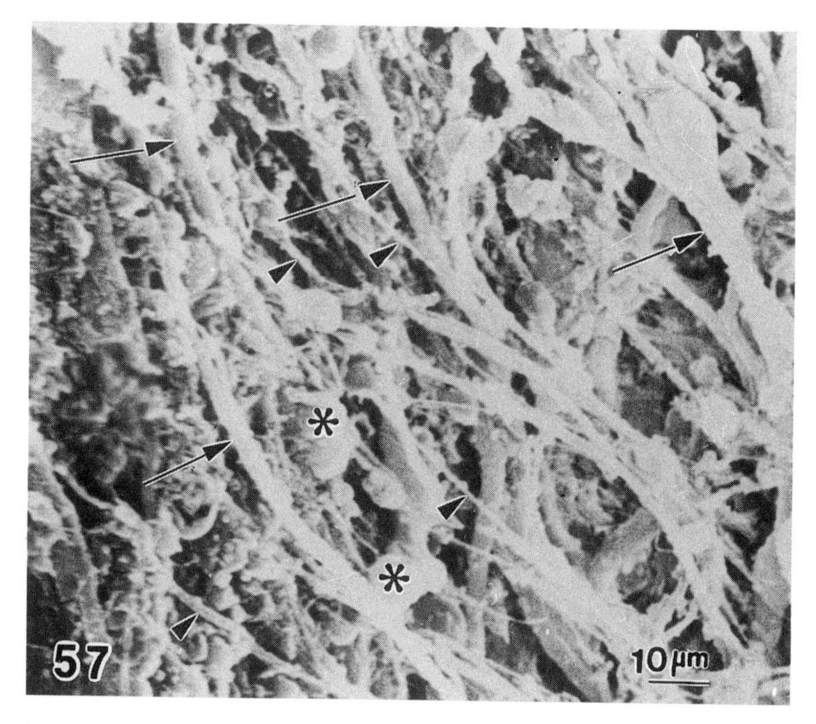

FIGURE 57. Low magnification scanning electron microscopy micrograph of teleost fish cerebellar white matter. The thick mossy fibers (arrows) are clearly distinguished from the thin climbing fibers (arrowheads). Oligodendrocytes (asterisks) are seen amongst the axonal bundles. SEM slicing technique. Gold–palladium coating.

Using the SEM slicing technique, climbing fibers in the teleost fish (Figure 58) are seen establishing synaptic contacts with the granule cell dendrites, thus forming small climbing fiber glomeruli (Castejón & Caraballo, 1980b). Fine tendril collaterals of climbing fibers are also observed spreading throughout the granular layer.

Climbing fibers in the teleost fish Purkinje cell layer are also traced with the SEM beam. The climbing fibers show an intimate topographic relationship with the Purkinje cell body (Figures 44–46). The freeze–fracture method SEM removed the satellite Bergmann cells and exposed the hidden outer surface of Purkinje cells. At the interface between the Purkinje and molecular layers, the climbing fiber collaterals ascend vertically and produce fine collaterals extending in three different planes in the molecular layer (Figure 59).

In the human cerebellar cortex processed by ethanol-cryofracturing technique (Figure 60), the cryodissected climbing fibers establish synaptic connections with Purkinje dendritic spines. In addition, some Scheibel's retrograde collaterals are seen descending to the granular layer (Castejón & Valero, 1980). In samples of teleost fish cerebellum, coated with gold–palladium and examined with SEM, the terminal endings of climbing fibers were found intimately applied to the outer surface of secondary and tertiary Purkinje cell

FIGURE 58. Teleost fish cerebellar granular layer. The climbing fibers (CF) are observed making synaptic contacts (arrowheads) with granule cell (gc) dendrites. The arrow indicates the tendril collaterals of climbing fibers. Scanning electron microscopy (SEM) slicing technique. Gold–palladium coating.

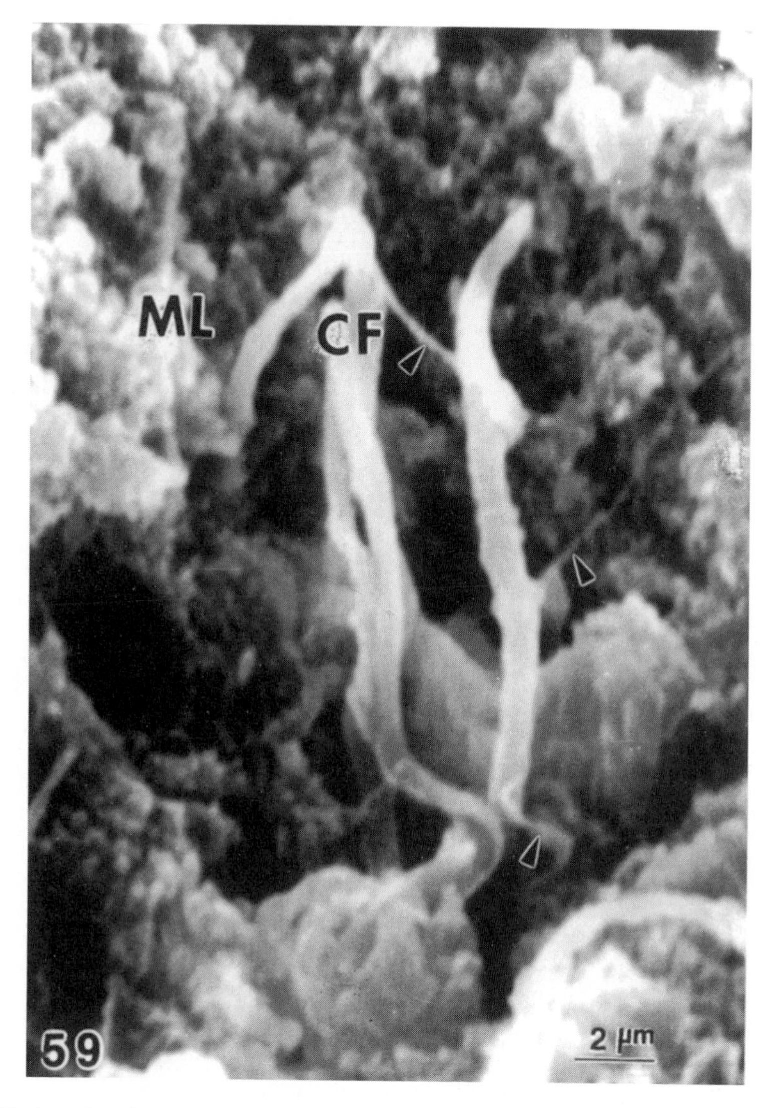

FIGURE 59. Scanning electron microscopy micrograph of teleost fish cerebellar cortex showing climbing fibers (CF) and their collateral processes (arrowheads) entering into the molecular layer (ML). Scanning electron microscopy (SEM) slicing technique. Gold–palladium coating.

dendrites (Figure 61) and making axospinous synaptic connections with Purkinje cell dendritic spine (Figure 62).

SEM examination of climbing fibers supports and extends previous light microscopic (LM) and TEM studies of climbing fibers (Castejón, 1986, 1988; Estable, 1923; Fox et al., 1967b; Hámori & Szentágothai, 1966a; Larramendi & Victor, 1967; Mugnaini, 1972; Palay & Chan-Palay, 1974; Ramón y Cajal, 1955; Scheibel & Scheibel, 1954).

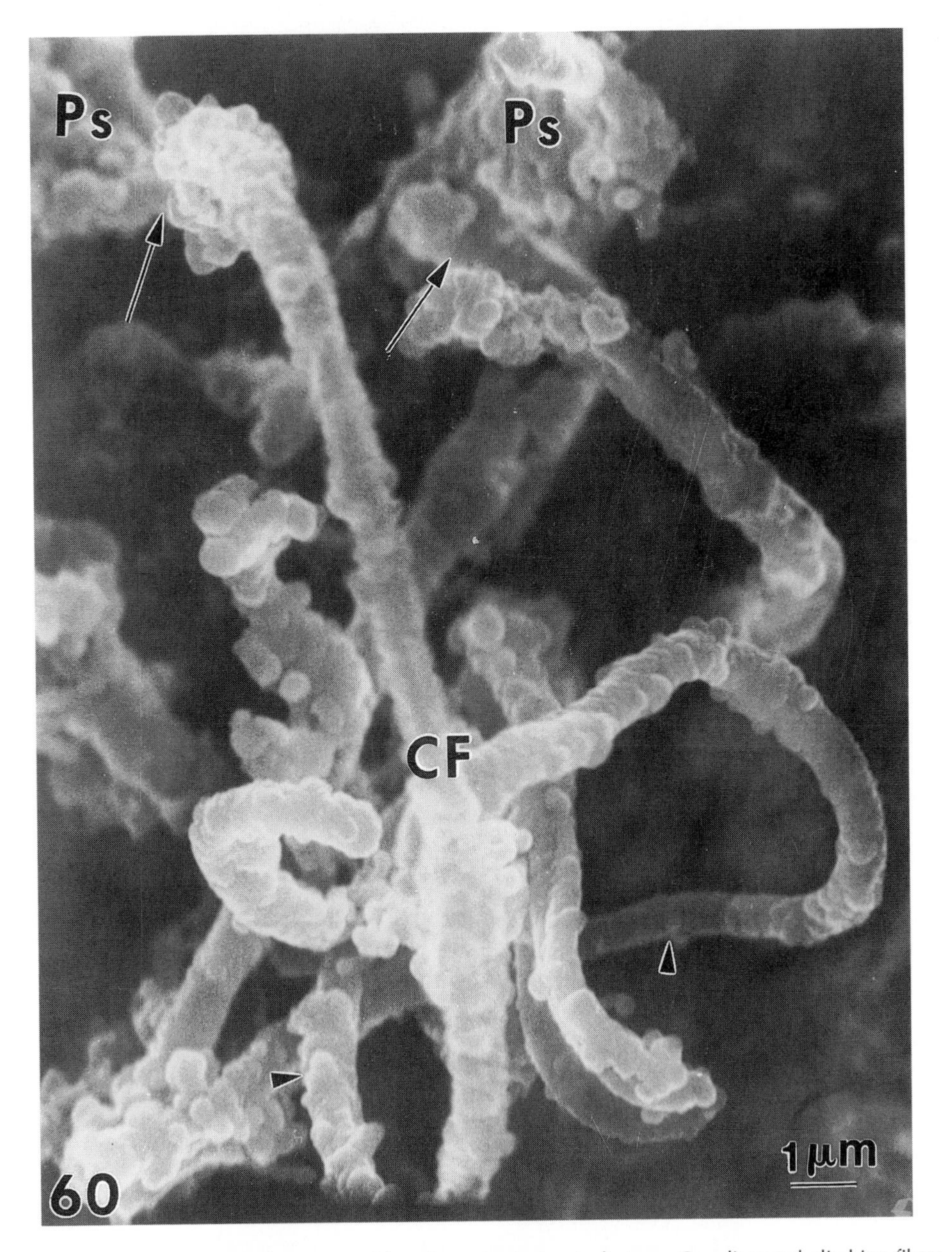

FIGURE 60. Human cerebellar cortex. Ethanol-cryofracturing technique. Cryodissected climbing fibers (CF) in the molecular layer demonstrating their typical crossing-over bifurcation pattern. The arrows identify the synaptic junctions with Purkinje dendritic spines (Ps). The arrowheads show the retrograde collaterals descending toward the granular layer. Gold–palladium coating.

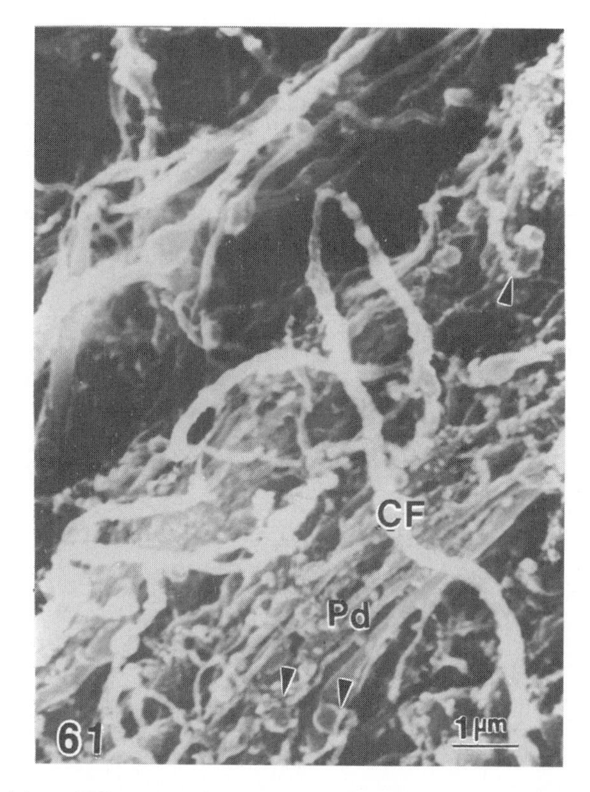

FIGURE 61. Teleost fish cerebellar cortex. Climbing fiber collaterals (CF) ending upon the surface of Purkinje tertiary dendritic ramifications (Pd). The arrowheads indicate the synaptic relationship with Purkinje dendritic spines. Freeze–fracture scanning electron microscopy (SEM) method. Gold–palladium coating.

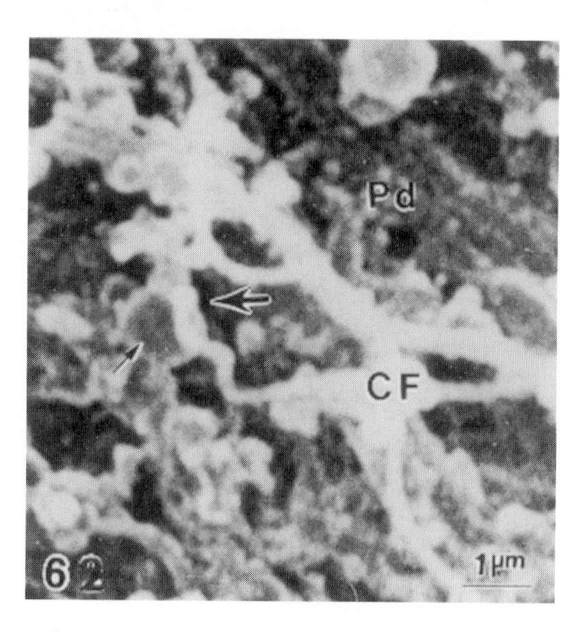

FIGURE 62. High magnification scanning electron microscopy micrograph of teleost fish cerebellar cortex. A climbing fiber collateral (CF) ends (thick arrow) upon a Purkinje dendritic (Pd) spine body (thin arrow). SEM freeze–fracture method. Gold–palladium coating.

CONCLUDING REMARKS

The slicing technique for SEM as well as the ethanol-cryofracturing technique and the freeze–fracture method for SEM have demonstrated, with more accuracy than other microscopical techniques, the high degree of lateral collateralization of climbing fibers in the granular and molecular layers and therefore their wide divergent field in the cerebellar cortex. The climbing fibers in these layers are not confined to a parasagittal and/or transverse plane, but instead dispersed in a wide radial fashion to a three-dimensional field. The climbing fiber terminals end as bulbous knobs making synaptic contacts with Purkinje dendritic spines.

Chapter 10

The Basket Cells

BRIEF HISTORY

Basket cells were first described by Golgi (1883) and Ramón y Cajal (1888). Ramón y Cajal called these cells "large or deep stellate cells." Kölliker (1890) referred to these cells as "basket cells" (*Korbzellen*). Ramón y Cajal (1888) traced the axonal plexus formed by basket cell axonal ramifications around the Purkinje cells, a discovery that led him to proclaim the Neuron Doctrine. Estable (1923) proposed designating the pericellular basket as *nid de Cajal.*

Scheibel and Scheibel (1954) reported that basket and stellate cells receive synaptic contacts from climbing fibers. Hámori and Szentágothai (1964) and Fox, Hillman, Siegesmund, and Dutta (1967a) demonstrated that parallel fibers make synaptic contacts with stellate and with basket cells. Hámori and Szentágothai (1965) gave a detailed description of basket cell–Purkinje cell synaptic relationship. Lemkey-Johnston and Larramendi (1968a, 1968b) published elegant electron microscopic studies of mouse stellate and basket cells. They established the criteria for sampling and classification of stellate and basket cells and also the quantitative distribution of types of nerve terminals upon basket and stellate cells. Monteiro (1989) reported the morphometric differences between basket and stellate cells. Gobel (1971) described the presence of axo–axonic septate junctions in the basket formation at the level of the *pinceaux*, at the initial segment of Purkinje cell axon. Palay and Chan-Palay (1974) gave the most impressive and detailed description of fine structural features of basket cells. Bishop (1993) and King, Chen, and Bishop (1993) reported the basket cell axonal distribution in the cat's cerebellum. Castejón, Castejón, and Sims (2001a) described a correlative microscopic study and the three-dimensional scanning electron microscopic (SEM) features of basket cells.

THREE-DIMENSIONAL MORPHOLOGY

The basket cells have a round, triangular, or oval shape and are localized in the lower third of the molecular layer or between the Purkinje cells with their long axis parallel to the Purkinje cell layer in the sagittal plane. The axon of basket cell emerges directly either from the cell body or from one of the major dendrites. It follows a horizontal course in the parasagittal plane running along several Purkinje cells. The axon gives off three types of collaterals: (1) descending collaterals that approximate in cascade over the primary dendritic trunk and cell bodies of several Purkinje cells, (2) ascending collaterals toward the molecular layer, and (3) transverse collaterals that remain in the plane of cell body and give off in turn, descending collaterals contributing to the Purkinje pericellular nest and *pinceaux* and also ascending collaterals directed to the molecular layer (Castejón et al., 2001a).

Scanning electron micrographs of human cerebellum processed by the ethanol-cryofracturing technique showed the axonal ramifications of basket cells embracing the Purkinje cell body (Figure 63). The ethanol-cryofracturing technique selectively removes the satellite Bergmann cell covering the Purkinje cell soma and exposes the basket cell axonal collaterals applied to the Purkinje cell outer surface soma (Figure 64). At higher

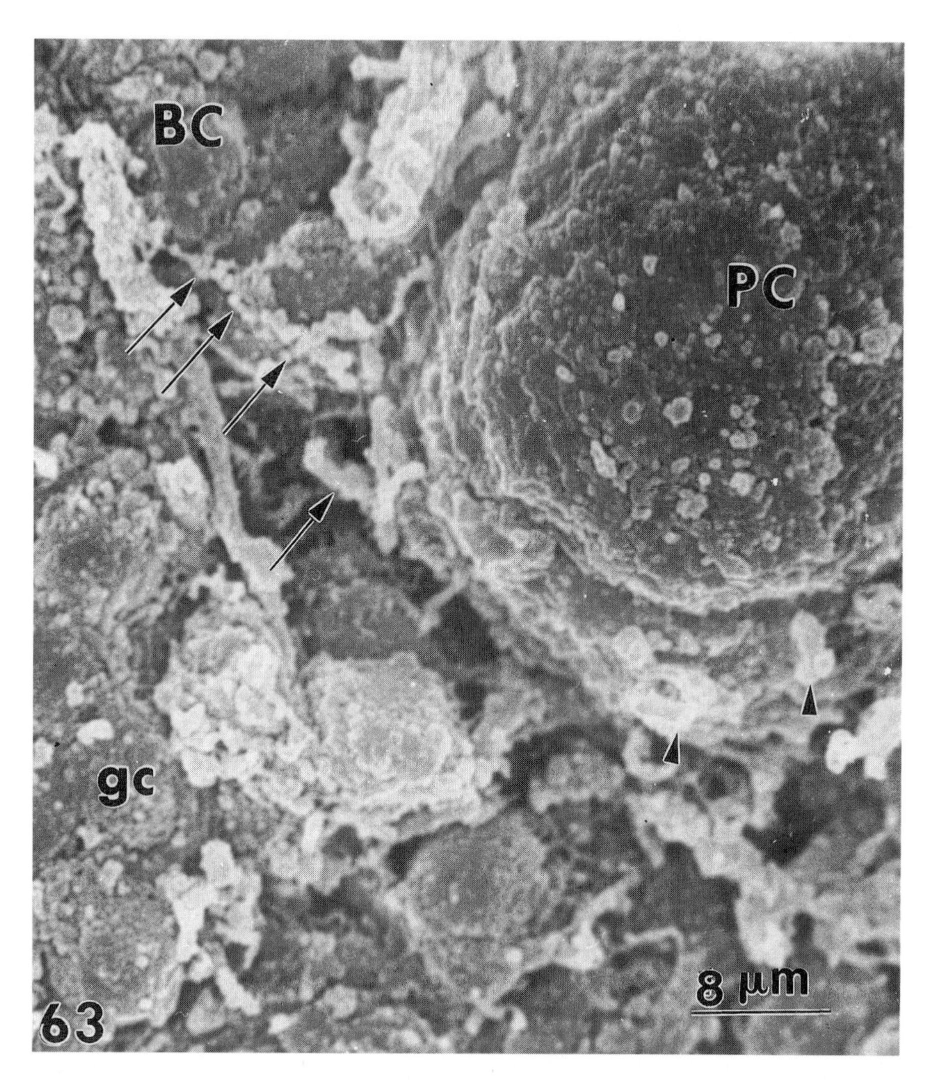

FIGURE 63. Scanning electron microscopy micrograph of human cerebellar cortex. Ethanol-cryofracturing technique. A basket cell (BC) located in the Purkinje cell layer is observed giving off its transverse axonal ramifications (arrows) toward a neighboring Purkinje cell soma (PC). The arrowheads show a partial view of Purkinje pericellular network formed by basket cell axonal endings embracing the Purkinje cell soma. The transversal plane of the fractograph allows also to observe the granule cells (gc). Gold–palladium coating.

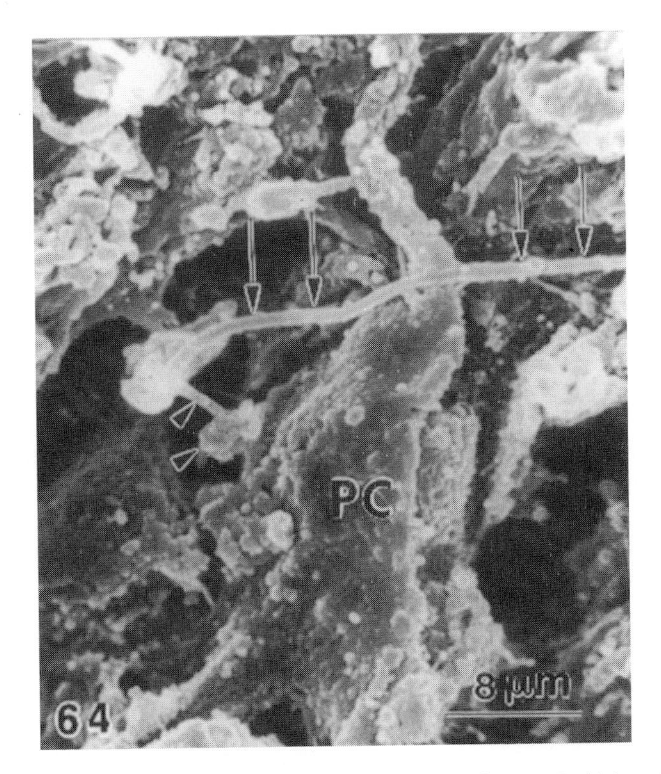

FIGURE 64. Scanning electron microscopy micrograph of the human Purkinje cell layer. Ethanol-cryofracturing technique. A transversal axonal process of a basket cell (arrows) is observed running across the Purkinje cell (PC) primary dendritic trunk and ending at the Purkinje cell soma (arrowheads), contributing to the formation of Purkinje pericellular nest. Gold–palladium coating.

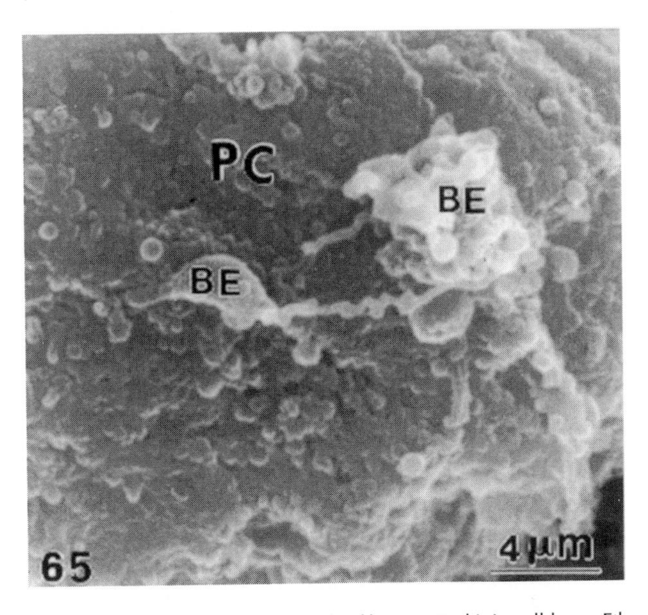

FIGURE 65. Scanning electron microscopy micrograph of human Purkinje cell layer. Ethanol-cryofracturing technique. Higher magnification showing round and oval interconnected basket cell axonal synaptic endings (BE) applied to the outer surface of Purkinje cell soma (PC). Gold–palladium coating.

magnification, the short varicose twigs formed by the basket cell endings appear to terminate as interconnected boutons (Figure 65). In the cerebellar cortex of teleost fish processed by the slicing technique, the axonal descending collaterals of basket cells are observed at the level of the Purkinje cell axonal initial segment contributing to the formation of the *pinceaux* (Figure 46).

CONCLUDING REMARKS

The ethanol-cryofracturing technique of Humphreys, Spurlock, and Johnston (1975) produces a cryodissection of the descending and transverse axonal collaterals of basket cells and a removal of the satellite Bergmann cell covering the Purkinje cell soma exposing the three-dimensional relief of the pericellular nest and the outer surface appearance of the *pinceaux*. Such SEM contribution to the study of this unique axosomatic synapse in the cerebellar cortex reveals the value of SEM for studying short intracortical circuits and synaptic morphology in the central nervous system.

Chapter 11

Stellate Cells

BRIEF HISTORY

Stellate cells, the intrinsic interneurons of the cerebellar molecular layer, were first described with light microscopy (LM) by Fusari (1883), Ramón y Cajal (1888, 1896), Ponti (1897), Smirnov (1897), Estable (1923), Jakob (1928), and Scheibel and Scheibel (1954). More recently, their fine structure was studied by Herndon (1964), Fox, Hillman, Siegesmund, and Dutta (1967a), Lemkey-Johnston and Larramendi (1968a, 1968b), Castejón (1968), Sotelo (1969, 1970), and Palkovitz, Magyar, and Szentágothai (1971). Chan-Palay and Palay (1972a) and Palay and Chan-Palay (1974) have given the most complete description of these neurons by means of *camera lucida* drawings of Golgi preparations and transmission electron microscopy (TEM). Castejón and Castejón (1997) reported the scanning electron microscopic (SEM) three-dimensional morphology, TEM freeze–fracture features and glycosaminoglycan histochemistry of stellate neurons.

THREE-DIMENSIONAL MORPHOLOGY

Fish cerebellar cortex specimens processed conventionally for SEM show superficial short-axon stellate neurons (Figure 66) with round, elliptical or fusiform somata in a parasagittal fracture of the outer third molecular layer. Since they are the only neurons in the upper molecular layer, these superficial stellate cells are easy to recognize; basket cells and few stellate cells also are found in the middle and inner thirds of the molecular layer. Numerous stellate cells are seen spreading throughout the outer third molecular layer. Dendrites with several beads radiate from the cell body toward neighboring Purkinje dendrites or stellate cells (Figure 67). The axon originates from of a typical triangular-shaped axon hillock and, after a short initial segment, bifurcates into tenuous varicose collaterals. Short, ramified, and beaded dendrites emerge from the cell somata, and are directed toward the passing bundles of parallel fibers.

Some larger stellate cells possess an axon that follows a winding course, exhibits short sectioned collateral branches, and round enlarged endings. Until now, neither the sections or fractures studied with the SEM allowed the observation of their ascending or descending

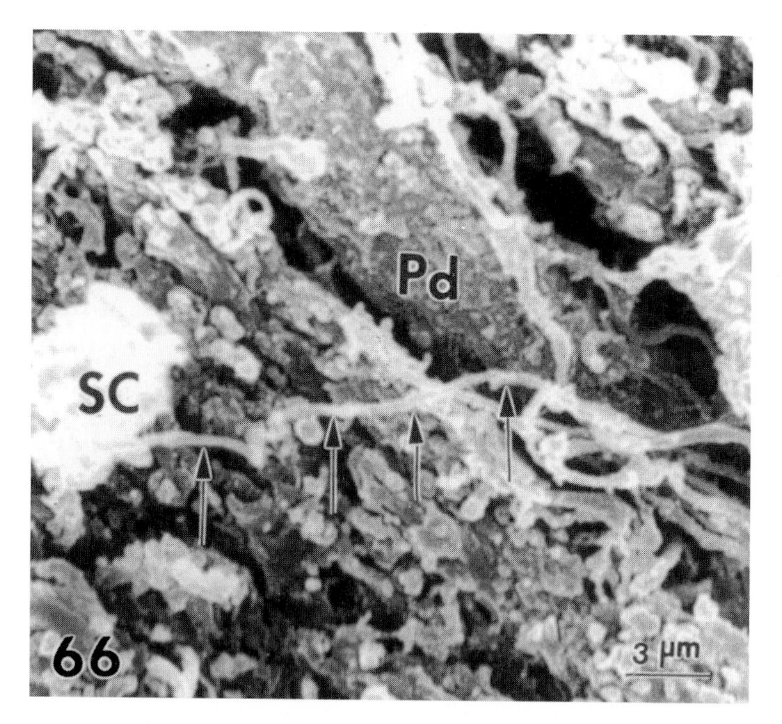

FIGURE 66. Scanning electron microscopy micrograph of human cerebellar molecular layer showing a stellate cell (SC) and its axonal process (arrows) directed toward a neighboring Purkinje dendrite (Pd). Ethanol-cryofracturing technique. Gold–palladium coating.

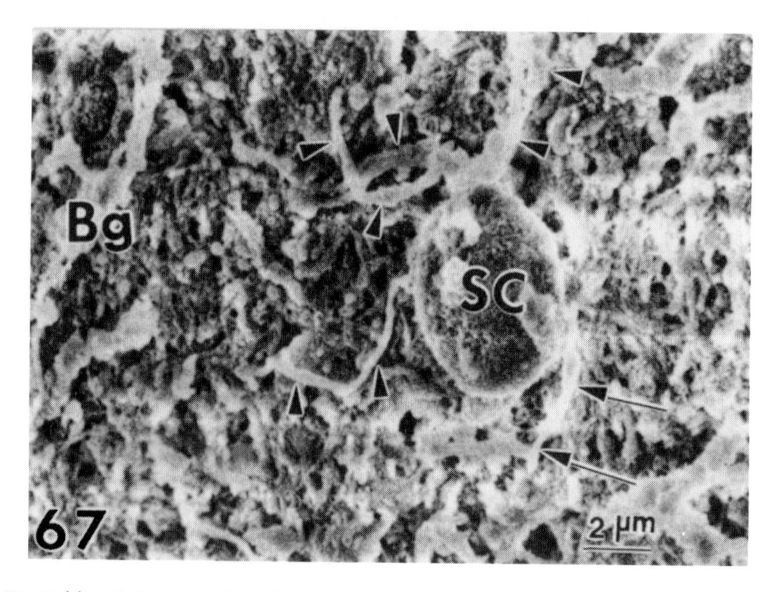

FIGURE 67. Field emission scanning electron microscopy micrograph of the outer surface of a stellate cell (SC) of mouse cerebellar molecular layer, processed by the freeze–fracture method. The round cell body shows a short axon (arrows) and three dendritic processes (arrowheads). A Bergmann fiber (Bg) also is distinguished. Chromium coating.

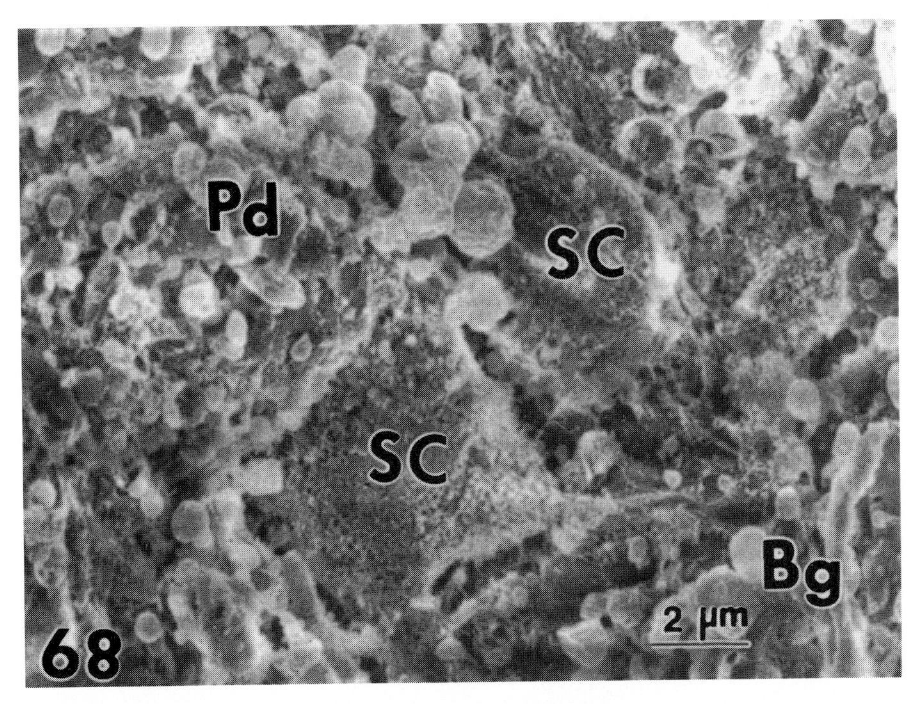

FIGURE 68. Field emission scanning electron microscopy micrograph of two fractured stellate cells (SC) of mouse cerebellar molecular layer surrounded by tertiary Purkinje spiny dendritic ramifications (Pd) and Bergmann fibers (Bg). SEM freeze–fracture method. Chromium coating.

collaterals, as classically described in Golgi LM impregnated material. The axonal terminal arborization appears as a delicate plexus that branches and rebranches over the Purkinje dendritic branchlets. Small varicosities could be seen along the course of the fine terminal axonal branches.

With the freeze–fracture technique for field emission SEM (FESEM), at low magnification, stellate neurons are seen fractured through the equatorial plane (Figure 68), and show a condensed pattern of nuclear heterochromatin and a thin rim of perinuclear cytoplasm.

These cells appear immersed within the bundles of parallel fibers and the Purkinje cell dendritic processes (Figure 69). At higher magnifications, and depending upon the plane of the cryofracture, the surfaces of the anastomotic bands of nuclear heterochromatin and endoplasmic reticulum of stellate neurons were clearly visualized, because of the "washing out" of soluble proteins from the nucleoplasm and cytoplasmic matrix induced by the thawing in absolute alcohol.

Stellate neurons fractured through the main axis of cell somata and dendritic processes show the three-dimensional image of the Golgi complex, endoplasmic reticulum and lysosomes (GERL) formed by the Golgi cisternae and their sacs, endoplasmic reticulum canaliculi and lysosomes (Figure 70).

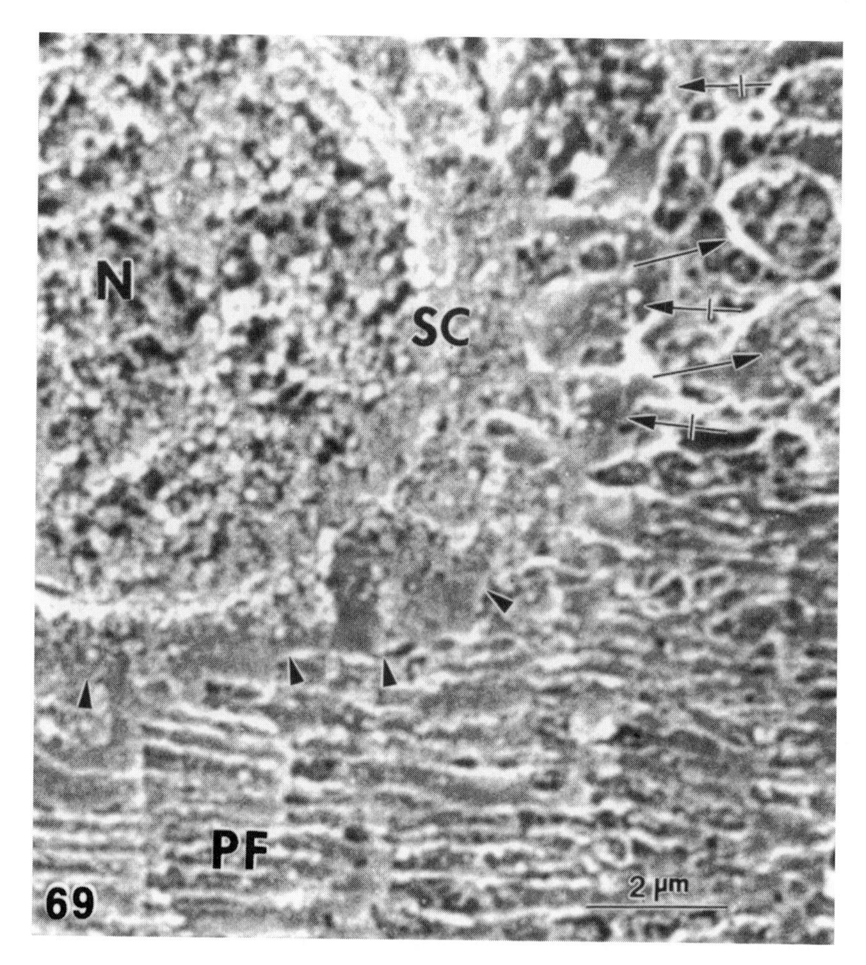

FIGURE 69. Teleost fish cerebellar molecular layer. A fractured stellate neuron (SC) is observed displaying the nucleus (N) and the scarce perinuclear cytoplasmic band (arrowheads). Note the neighboring longitudinal course of non-synaptic segments of parallel fibers (PF) and the cross-fractured synaptic varicosities of parallel fibers (arrows) containing spheroid synaptic vesicles. Some synaptic endings, presumably belonging to climbing fiber collaterals and containing spheroid synaptic vesicles, establish axosomatic contacts (cross arrows) with the stellate cell plasma membrane. SEM freeze–fracture method. Gold–palladium coating.

CONCLUDING REMARKS

Since the stellate cells are the only short axon nerve cells existing in the outer third molecular layer with several circumscribed contorted dendrites, these microneurons are easily identified by SEM. The stellate cell axon appears as a unique process showing a typical axon hillock at the initial segment and terminal ramification at the sites of contact with Purkinje dendrites. The contorted stellate dendrites exhibit a beaded aspect and frequent bifurcations. The low resolution of SEM does not allow us until now to

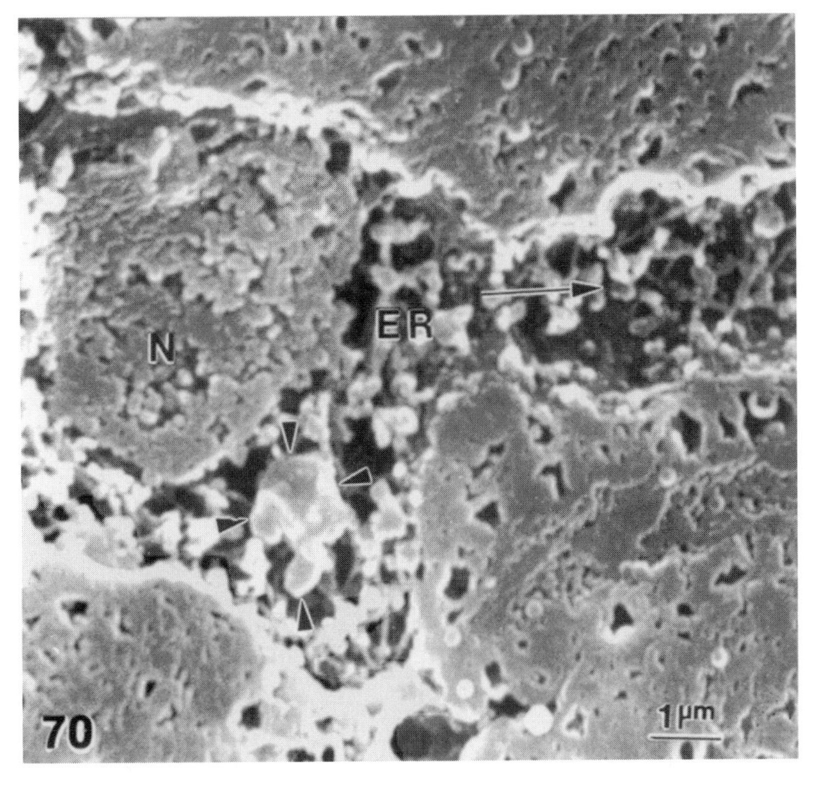

FIGURE 70. Fractured stellate cell of teleost fish cerebellar molecular layer showing the fractured nucleus (N), the Golgi complex (arrowheads) and the endoplasmic reticulum profiles (ER) extended from the cell soma to the dendritic process (arrow). SEM freeze–fracture method. Gold–palladium coating.

characterize the synaptic relationship of stellate cell axons with Purkinje cell dendrites and the stellate cell dendrite-parallel fiber synapses as formerly demonstrated by TEM (Castejón, 1968; Palay & Chan-Palay, 1974). Some internal details of fractured stellate neurons have been three-dimensionally viewed by freeze–fracture SEM, taking advantage of the "washing out" of soluble proteins from the fracture face induced by the freeze–fracture process (Haggis, Bond, & Phipps-Todd, 1976).

Chapter 12

Cerebellar Glial Cells

BRIEF HISTORY

Cerebellar neuroglial cells were initially described by Bergmann (1857), Golgi (1885), Gierke (1886), Van Gehuchten (1891), Retzius (1892c), Weigert (1895), Ramón y Cajal (1896a), Terrazas (1897), Fañanás and Ramón y Cajal (1916), Jakob (1928), Schroeder (1929), and Jansen and Brodal (1958). More recently, a complete account of transmission electron microscopic (TEM) features and stereograms of Golgi preparations, performed by high voltage electron microscopy, was given by Palay and Chan-Palay (1974). Hanke and Reichenbach (1987) and Siegel et al. (1991), using the rapid Golgi technique, described the Bergmann cells in the cerebella of rat of various ages, measured the length and diameter of Bergmann fibers, and quantified the presence of bushy lateral protrusions. Suarez et al. (1992) reported, by means of immunocytochemical methods, a different response of astrocytes and Bergmann cells to portocaval shunt.

Bergmann cells of several vertebrates also were examined by means of freeze–fracture and ethanol-cryofracturing techniques for conventional scanning electron microscopy (SEM) and TEM to study their three-dimensional morphology and topographic relationships in the molecular layer (Castejón, 1990b). Bergmann cell morphology also has been demonstrated by means of chromogranin A-like immunoreactivity (McAuliffe & Hess, 1990). Reichenbach et al. (1995) applied immunocytochemistry for glial fibrillary acidic protein (GFAP) and glutamine synthetase, to selectively label Bergmann cells in adult cerebellum and studied the distribution of their somata and processes. Kugler and Drenckhahn (1996) reported that Bergmann cells are important sites of nitric oxide synthase I suggesting that these cells might use nitric oxide as gaseous messenger molecule for various aspects of glia-neuron signaling. Muller, Moller, Neuhaus, and Kettenmann (1996) showed in mouse cerebellar cortex electrical coupling among Bergmann cells and their modulation by glutamate receptor activation using Lucifer Yellow injection, patch-clamp cell pairs, and ultrastructural examination. Kril, Flowers, and Butterworth (1997) reported increased GFAP immunoreactivity of Bergmann cells in human hepatic encephalopathy. Lafarga, Andres, Calle, and Berciano (1998) described phagosomes containing apoptotic bodies in reactive immature Bergmann cells treated with methylazoxymethanol (MAM).

OLIGODENDROCYTES

These cells mainly are encountered in the white matter but also occur in the granular, Purkinje cell, and molecular layers. They are associated with axonal afferent mossy and climbing fibers, and are intercalated between granule cells and as satellites of Purkinje cells. Oligodendrocytes in perivascular position are also found. Figure 71 shows oligodendrocytes of the cerebellar white matter, in topographic relationship with mossy and climbing fibers. These glial cells are round or oval cells, 5–6 μm in diameter, exhibiting mainly two processes. In the teleost fish granular layer, they appear as oval cells located among the collaterals of climbing and mossy fibers and around the granule cell bodies (Figure 72).

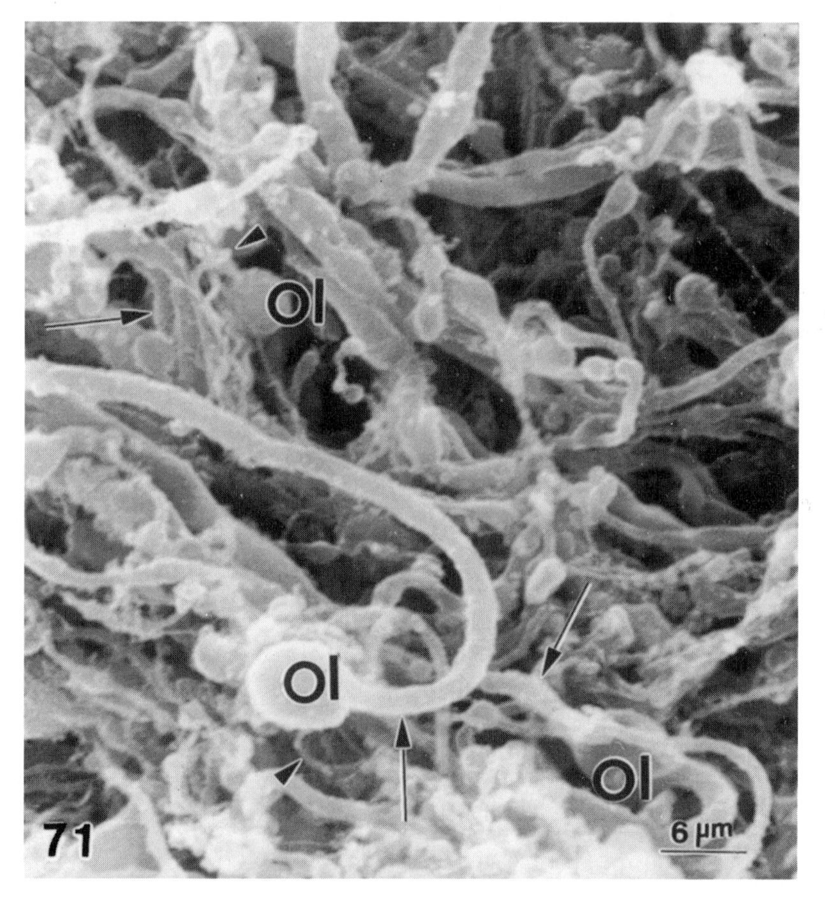

FIGURE 71. Scanning electron microscopy micrograph of teleost fish cerebellum. White matter showing the cerebellar afferent fibers (arrows) and their associated oligodendrocytes (OL). These cells exhibit two or three processes (arrowheads) extending to different axons. SEM slicing technique. Gold–palladium coating.

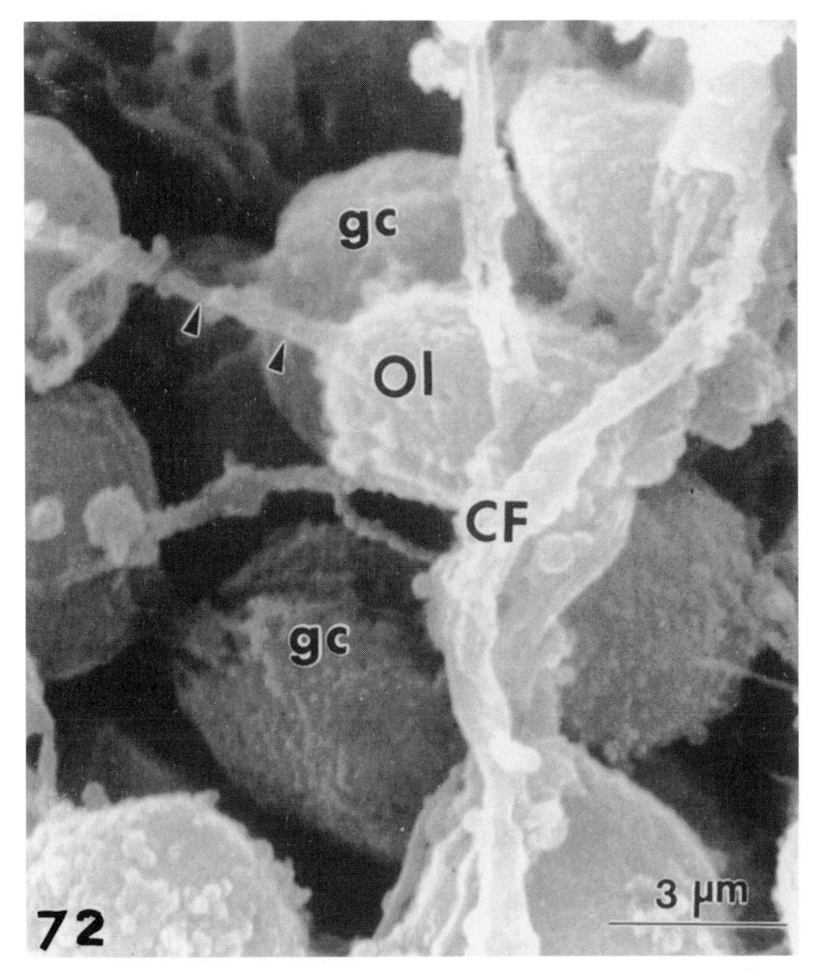

FIGURE 72. Scanning electron microscopy micrograph of teleost fish cerebellar granular layer. An oligodendrocyte (Ol) and its process (arrowheads) is observed between a climbing fiber (CF) and the granule cells (gc). SEM slicing technique. Gold–palladium coating.

Oligodendrocytes examined by the freeze–fracture method for SEM (Figure 73) show the inner surface details of nuclear chromatin, nuclear envelope, and the three-dimensional and spatial distribution of endoplasmic reticulum canaliculi and vesicles. Microtubules and suspended cell organelles, such as mitochondria and lysosomes, are also visualized. A nuclear fibrillar inclusion is also present.

Microglial Cells

Although we have systematically examined normal cerebellar samples in most vertebrate species, we have been unable to rationally characterize thus far the microglial cells by SEM.

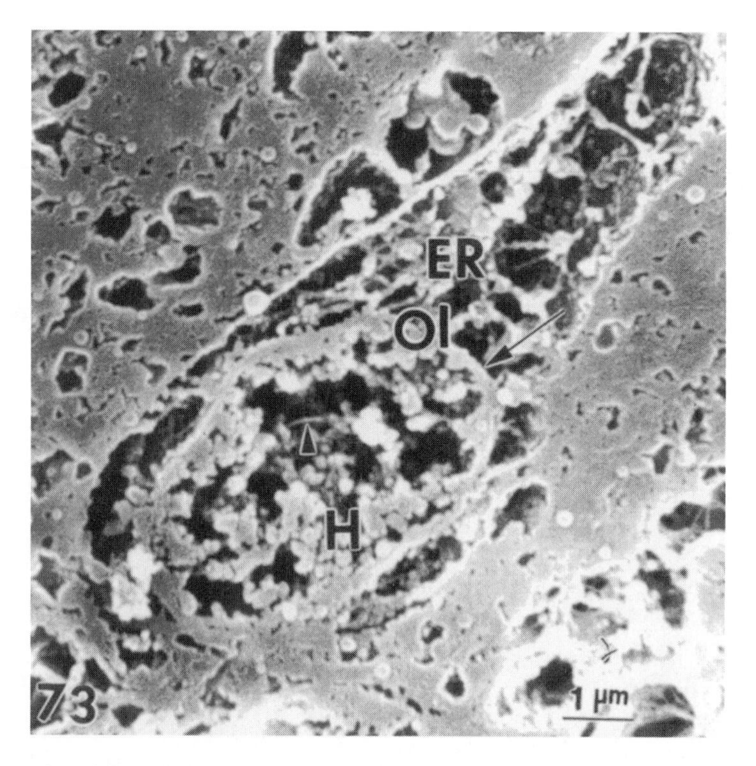

FIGURE 73. Teleost fish cerebellum processed by the freeze–fracture scanning electron microscopy (SEM) method. A fractured oligodendrocyte (Ol) exhibits the inner nuclear and cytoplasmic surfaces. Note the nuclear heterochromatin masses (H), a fibrillar intranuclear inclusion (arrowhead), the nuclear envelope (arrow), and the stereo-spatial distribution of endoplasmic reticulum (ER). Gold–palladium coating.

THE VELATE PROTOPLASMIC ASTROCYTES

These types of protoplasmic astrocytes were first described by Chan-Palay and Palay (1972b) and Palay and Chan-Palay (1974) using the rapid Golgi method for light microscopy (LM) and high voltage electron microscope (EM). They have also been characterized using field emission SEM (FESEM) (Castejón, Castejón, & Apkarian, 2001c). They have been preserved using delicate handling of the tissue and chromium coating and are found in the granular layer, typically characterized by their laminar processes which wrap, almost completely, the granule and Golgi cells and glomerular regions. They appear like a veil, forming a pericellular network that divides the granular layer into neighboring compartments (Figure 74). These velamentous walls, composed of neuroglial processes, seem to be part of the morphological substrate of modular arrangement of cerebellar granular layer. These astrocytic laminar processes show a high type II secondary electron (SE-II) topographic contrast when compared with that of granule cells. The extensive laminar processes can also be seen in relationship with cerebellar capillaries.

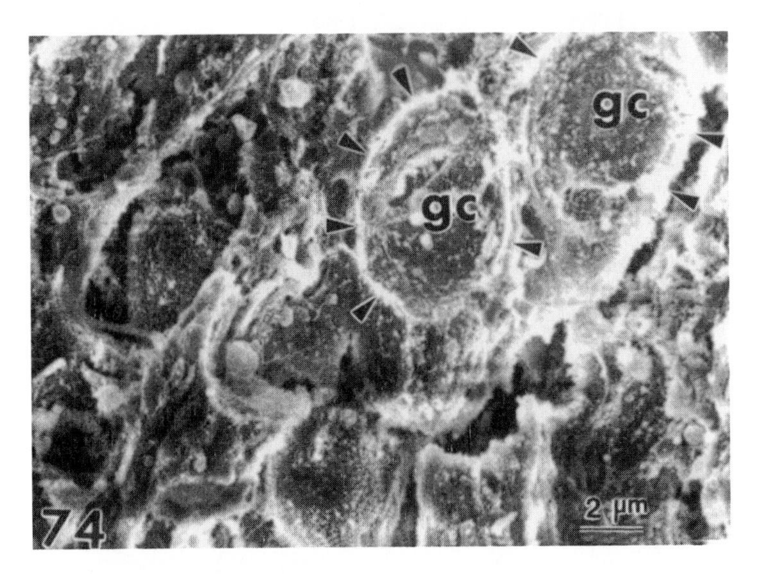

FIGURE 74. Field emission scanning electron microscopy micrograph of fractured mouse granular layer showing the high mass density processes of velate protoplasmic astrocytes (arrowheads) surrounding the fractured granule cells (gc). Freeze–fracture method. Chromium coating.

BERGMANN GLIAL CELLS

With conventional SEM of teleost fish cerebellar cortex, these cerebellar regional astrocytes appear as round, elongated, or triangular cells closely applied to the Purkinje cell soma and dendritic ramifications (Figure 75). At the level of FESEM, the Bergmann cells show a smooth, high mass density outer surface with numerous folial expansions that penetrate deep in the molecular layer (Figure 76). They surround the Purkinje cell soma and dendritic arborization and form a continuous sheet that covers parallel fibers, basket, and stellate cells, and the dendrites of Golgi and Lugaro cells that ascend into the molecular layer (Figure 77). This complete sheet is formed by the imbrication of lamellae from several Bergmann glial cells. These observations fully support the description made by Palay and Chan-Palay (1974) in samples stained by the rapid Golgi method for LM and high voltage EM. At the level of parallel fiber bundles in the molecular layer, Bergmann cells are attached to the outer surface of non-synaptic segments of parallel fibers (Figure 78).

By high magnification, with a high resolution FESEM, the smooth outer surface of Bergmann cells appears to ensheath the synaptic varicosities of parallel fibers which contain spheroid synaptic vesicles (Figure 79). The vertical ascending processes of Bergmann cells or Bergmann fibers, emerge from the upper pole of the cell, and ascend straight through the entire molecular layer toward the pial surface. At FESEM level, Bergmann fibers are seen ascending to the pial surface where they end as globular expansions. They exhibit numerous collateral processes that form an intricate and dense supporting network in the molecular layer (Figure 80).

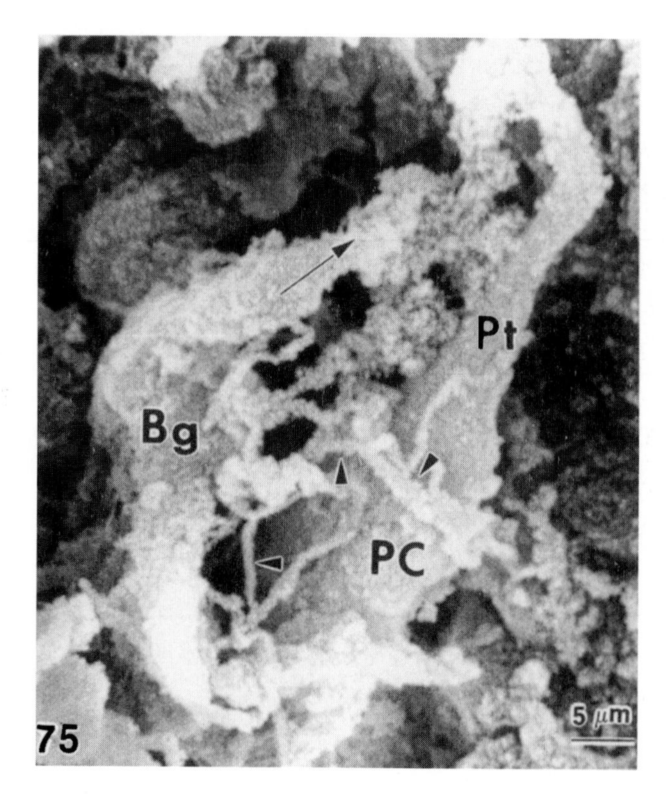

FIGURE 75. Human cerebellum. Ethanol-cryofracturing technique. The Bergmann cell (Bg) appears as a satellite of a Purkinje cell (PC). They are separated by the processes that form the pericellular nest (arrowheads) of Purkinje cell. The Bergmann fiber (arrow) originates from the upper pole of the Bergmann cell and is applied to the primary dendritic trunk (Pt) of Purkinje cell. Scanning electron microscopic (SEM) slicing technique. Gold–palladium coating.

FIGURE 76. Field emission scanning electron microscopy micrograph of mouse cerebellar cortex showing the outer surface of a Bergmann cell (Bg) apposed to the Purkinje cell soma (PC). A stellate cell (SC) shows its axonal process (arrow) directed to a Purkinje spiny dendritic branch (Pd) ensheathed by the laminar processes of Bergmann cell. SEM freeze–fracture method. Chromium coating.

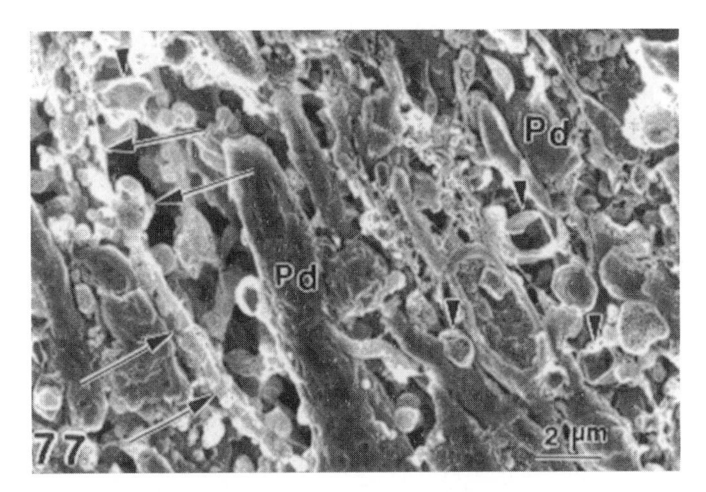

FIGURE 77. Field emission scanning electron microscopy micrograph of mouse cerebellar molecular layer. The veil processes of Bergmann cells (arrows) ensheath the spiny Purkinje dendritic ramifications (Pd) and also the cross-fractured non-synaptic segment of parallel fibers (arrowheads). SEM freeze–fracture method. Chromium coating.

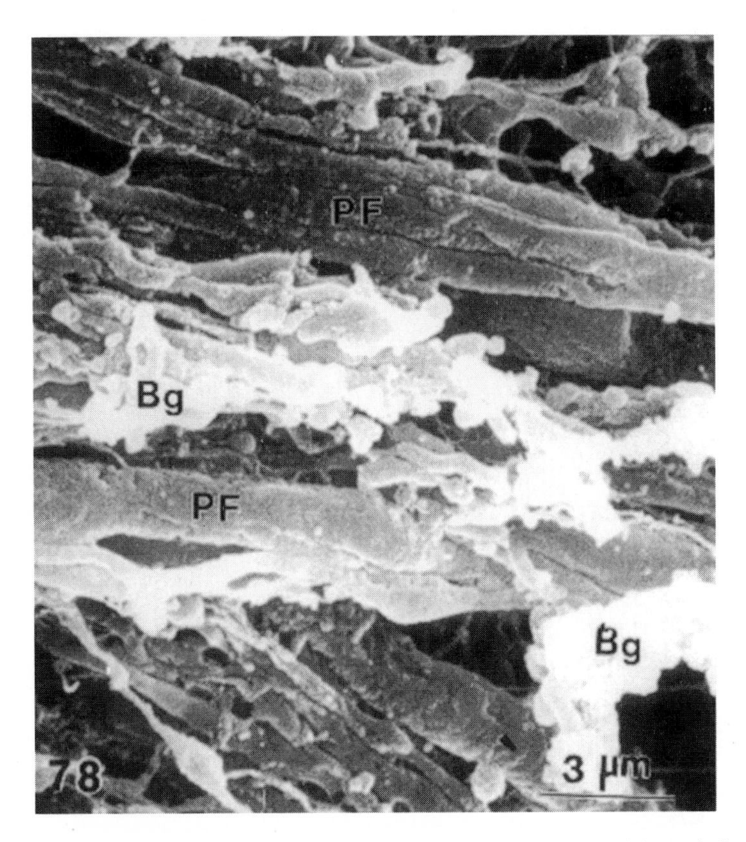

FIGURE 78. Conventional scanning electron microscopy micrograph of teleost fish cerebellar molecular layer showing the Bergmann cell cytoplasm (Bg) attached to the outer surface of the non-synaptic segments of parallel fibers (PF). The freeze–fracture method exposed the longitudinal course of parallel fiber bundles. Gold–palladium coating.

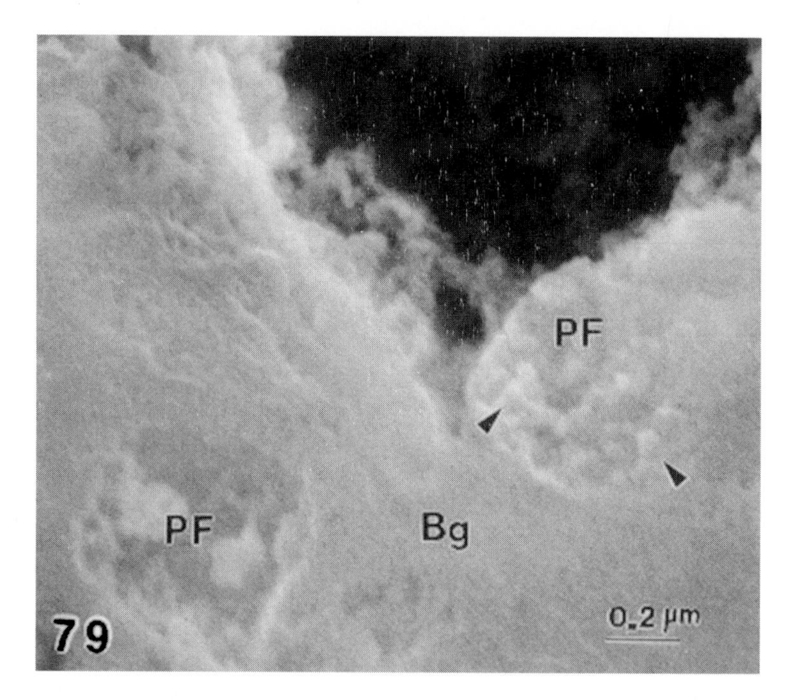

FIGURE 79. High magnification field emission scanning electron microscopy micrograph of primate cerebellar molecular layer. The cotton-like appearance and smooth surface of Bergmann cell cytoplasm (Bg) ensheathes the synaptic varicosities of two parallel fibers (PF) containing spheroid synaptic vesicles (arrowheads). SEM freeze–fracture method. Chromium coating.

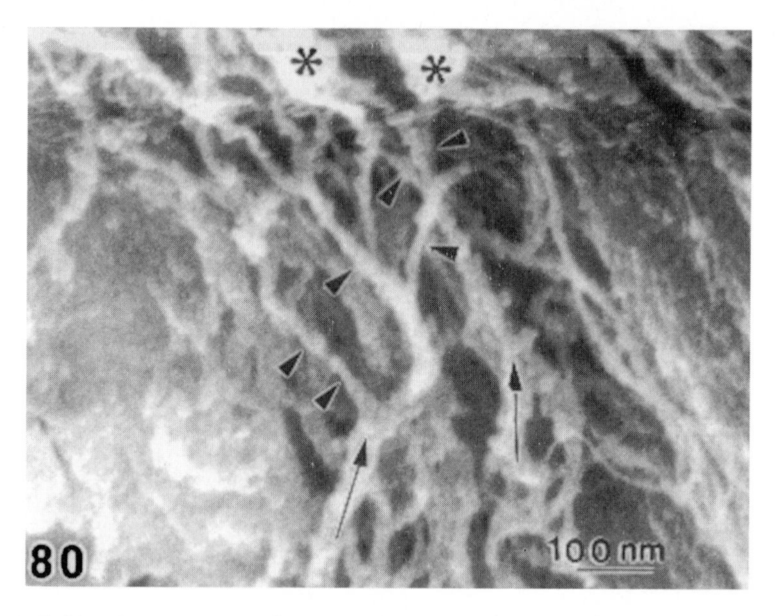

FIGURE 80. Field emission scanning electron microscopy micrograph of mouse cerebellar molecular layer showing the Bergmann fibers (arrows) and their collateral processes (arrowheads) ascending in the outer third molecular layer and ending as globular expansions (asterisks) at the level of the *glia limitans*. The depth of focus of SEM allows observation of the dense glial network formed by Bergmann fibers in the molecular layer. SEM freeze–fracture method. Chromium coating.

CONCLUDING REMARKS

Oligodendrocytes were characterized as round or oval cells closely associated with afferent mossy and climbing fibers. Velate protoplasmic astrocytes appeared as high mass dense cells with laminar processes that wrap granule and Golgi cells as well as glomerular regions. The cryofracture SEM method, the chromium coating technique and the FESEM made possible the visualization of the high mass, dense cytoplasm of Bergmann cell ensheathing the Purkinje cell soma and also the dendritic ramifications of all nerve cells and processes in the molecular layer. FESEM offered a new SE-I cotton-like image, with high topographic contrast of Bergmann cells and supported their intimate relationships at the Purkinje cell and molecular layers as previously demonstrated by TEM techniques (Castejón, 1990b).

Chapter 13

Cerebellar Capillaries

SHORT HISTORY

Most knowledge on scanning electron microscopy (SEM) of endothelial cells comes from studies using normal endothelium of large arteries, veins, and sinusoids (Bauman, Imparto, Kim, Yoder, & Grover-Thonson, 1978; Becker & De Bruyn, 1975; Buss, Klose, & Hollweg, 1976; Cho & De Bruyn, 1979; Christensen & Garbasch, 1972; Davies & Bowyer, 1975; Edanaga, 1974; Gertz, Rennels, Forbes, & Nelson, 1975; Groniowsky, Wiczyskowa, & Walski, 1971; Kawamura, Gertz, Sunaga, Rennels, & Nelsen, 1974; Lee & Chieng, 1979; Matonoha & Zechmaister, 1978; Nousek-Goebel & Press, 1986; Paine & Low, 1975; Shimamoto, Yamashita, & Sunaga, 1969; Stewart, Ritchie, & Lynck, 1973; Sunaga, Shimamoto, & Nelson, 1973; Swinehart, Pently, & Kardong, 1976; Wheeler, Gavin, & Herdson, 1973; Wolinsky, 1972). These earlier papers have examined endothelial surface morphological characteristics as seen with immersion and perfusion fixation techniques, and studied their changes related to the effect of pressure on the luminal surface, regional permeability, potential sources of artifacts as well as functional and pathological aspects.

Connections of the cerebellar granule cells with capillaries were first observed by Valenzuela-Chacón (1970) in the cerebellar cortex. The ultrastructure of cerebellar capillaries was further described by Lange and Halata (1972, 1979). Later, Pastukhov (1974), and Vaquero-Crespo (1975) described the interrelations of capillaries and cerebellar neurons. Del Cerro (1974) reported uptake of tracer proteins by blood vessels in the developing cerebellum. Hirano, Cervos-Navarro, and Ohsugi (1976) described the fine structure of the small vessels in the subarachnoid space of rat and mouse cerebellar cortex. Lange (1977) performed a comparative study of cell and vascular density in the cerebellar cortex. Larina (1980) described certain features of the capillary network of dog cerebellar cortex. Kotskovich (1981) studied the interactions of Purkinje cells, capillaries, and glial cells in cat cerebellar cortex. Duvernoy, Deon, and Vannson (1983) examined the vascularization of human cerebellar cortex. Heinsen and Heinsen (1983) reported qualitative and quantitative observations on cerebellar capillaries.

A correlative microscopic study of teleost cerebellar capillaries was reported earlier (Castejón, 1983a). The freeze–fracture method for SEM (Castejón, 1981; Castejón & Caraballo, 1980a, 1980b; Castejón & Valero, 1980; Haggis & Phipps-Todd, 1977;

Humphreys, Spurlock, & Johnston, 1974) was applied for the study of cerebellar capillaries. This method allowed the simultaneous visualization of the endothelial cell luminal surface, and also the cytoplasmic details of endothelial, pericyte, and astrocyte cells. This provided a new and distinctive range of observations and further topographic image capability of SEM to study the structural basis of capillary permeability.

The developmental and mature microvascular architecture of the rat and turtle cerebellar cortex have also been studied by using a cerebrovascular casting method for SEM observations (Akima, Nonaka, Kagesewa, & Tanaka, 1987; Kleiter & Lametschwandtner, 1995; Yoshida, Ikuta, Watabe, & Nagata, 1985). Sprouting endothelial cells containing cytoskeletal microtubules and microfilaments were reported in neonatal rat cerebellar cortex (Nousek-Goebel & Press, 1986). The inside-out capillary branching pattern has also been described in the three-layered structure of the cerebellar cortex (Yu, Yu, & Robertson, 1994). Recently, the formation of new capillaries following chronic hypoxia in the cerebellar granular layer was described by Bovero, Ascher, Arregui, Rovainen, and Woolsey (1999).

THREE-DIMENSIONAL MORPHOLOGY OF CEREBELLAR CAPILLARIES

The cross- and longitudinal sections of cerebellar capillaries have been studied in the granular layer, fibrous stratum, Purkinje cell, and molecular layers of teleost cerebellar cortex. In the granular layer, the capillaries pass between the glomerular regions and the granule cell groups. The perivascular neuropil consists mainly of myelinated axons, dendritic processes, oligodendrocytes, and perivascular astrocytes. Capillaries are also observed surrounding the Golgi cells in a satellite or perineuronal localization. The capillaries pierce the fibrous stratum (Castejón & Caraballo, 1980b) and reach the Purkinje cell and molecular layers. Capillaries usually bifurcate at the fibrous stratum and follow a vertical pathway to the molecular layer.

Teleost cerebellar capillaries are similar to invertebrate and vertebrate brain capillaries (Barber & Graziadei, 1967; Bruns & Palade, 1968; Castejón, 1980) and consist of a triple layered structure: (1) the endothelial cell layer; (2) the basement membrane with the enclosed pericytes; and (3) the glial cell layer formed by the perivascular astrocyte endfeet. According to Simionescu, Simionescu, and Palade (1974) and other studies carried out in human brain capillaries (Castejón, 1980), the endothelial cell can be divided into four well-defined zones for the purpose of its systematic study: the nuclear zone, the organelle zone, the peripheral cytoplasmic zone, and the endothelial junction zone.

With SEM higher magnification, a three-dimensional view of cytoplasmic details of endothelial cell zones and perivascular astrocytes can be gained in properly oriented fractographs (Figure 81). In sagittal sections, the endothelial nuclear zone showed a round prominent nucleus with coarse granulated heterochromatin and round, anastomotic profiles of endoplasmic reticulum of the abluminal cytoplasm. The endothelial organelle zone appears as a perinuclear triangular-shaped area with its base orientated toward the nuclear zone and its vertex continuous with the peripheral cytoplasm.

At the level of the nuclear zone, the endothelial cytoplasm forms a thin band surrounding the cell nucleus and exhibits free dense vacuoles and surface connected micropinocytotic vesicles. Numerous grey globular structures presumably lipoproteins,

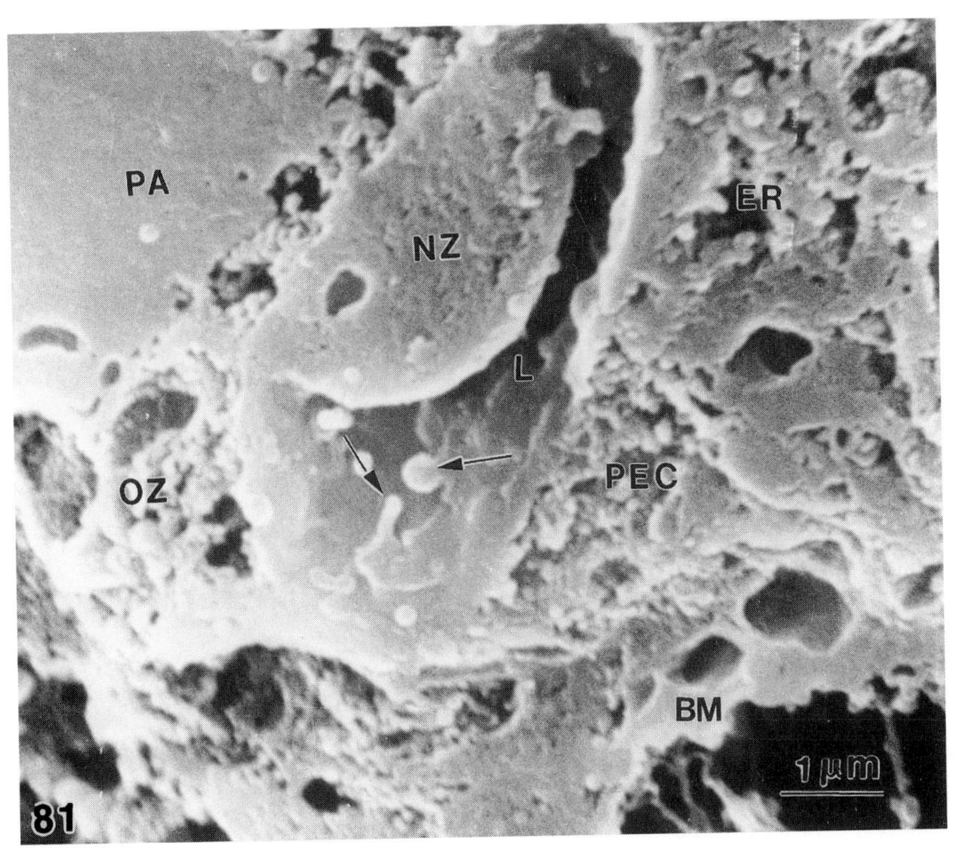

FIGURE 81. Liquid nitrogen frozen teleost fish *Arius spixii* cerebellar cortex. Scanning electron microscopy fractograph of a capillary cross-section showing the fractured endothelial cell nuclear (NZ) and organelle zones (OZ). The fractured peripheral endothelial cytoplasm (PEC) displays endoplasmic reticulum profiles (ER). The capillary lumen (L) and the endothelial microvilli (arrows) are seen. The perivascular astrocytic cytoplasm (PA) shows a typical glassy surface appearance. A segment of the capillary basement membrane (BM) is also noted. Scanning electron microscopy (SEM) freeze–fracture method. Gold–palladium coating.

occur in the capillary lumen as constitutive structural units of plasma substance. The peripheral cytoplasm exhibits clear or dense, fused or chained micropinocytotic vesicles almost extending between the luminal surface and the basement membrane.

SEM fractographs of sagittal sections of endothelial nuclear and organelle zones (Figures 82 and 83) exhibit deep nuclear invaginations containing the plasma globular structures. These structures were also observed randomly distributed throughout the fractured surface. The organelle zone offers a three-dimensional view of interconnected endoplasmic reticulum canaliculi, vesicles, and cisternae.

The SEM fractographs of peripheral endothelial cytoplasm obtained from fractures passing through the capillary longitudinal axis show the spherical, bright, surface connected micropinocytotic vesicles. Also, the isolated free micropinocytotic vesicles migrating through the smooth surfaced endothelial cytoplasmic matrix, and the dark,

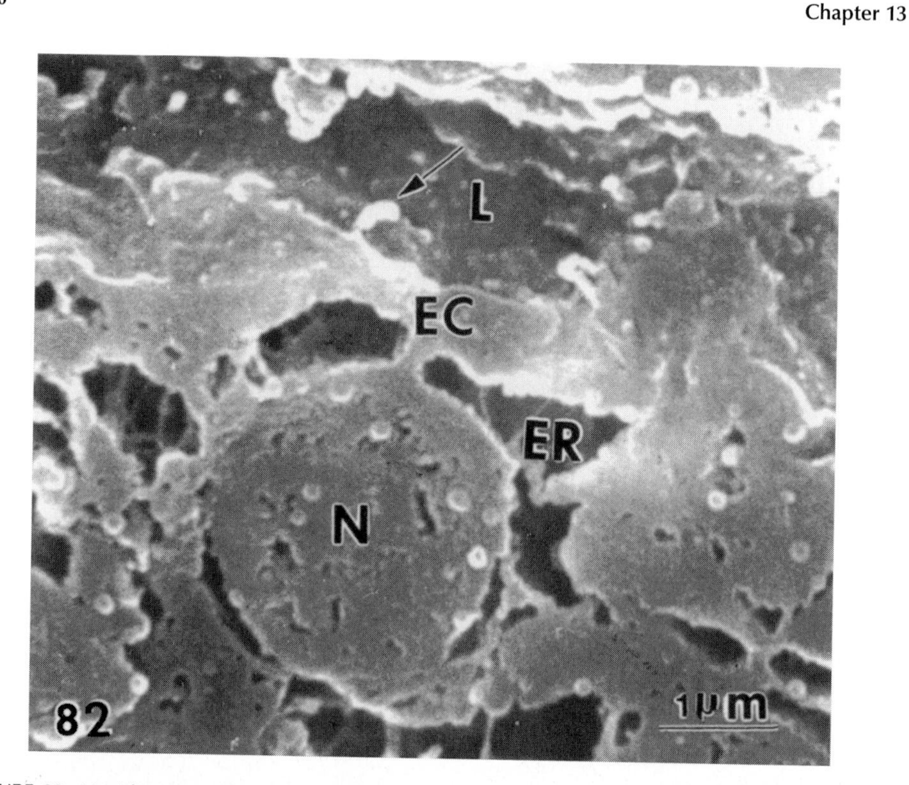

FIGURE 82. Liquid nitrogen frozen teleost fish *Arius spixii* cerebellar cortex. Longitudinally fractured capillary showing the capillary lumen (L) and the endothelial cell (EC) exhibiting the nucleus (N). The endoplasmic reticulum (ER) cisternae and microvilli (arrow) are also distinguished. Scanning electron microscopy (SEM) freeze–fracture method. Gold–palladium coating.

chained micropinocytotic vesicles forming fused or shuttled vesicles are seen (Figure 84). The vacuoles appear as dark structures, bound by a brilliant limiting membrane, with fused micropinocytotic vesicles. The rough endoplasmic reticulum appears as dark cored elongated profiles containing spherical particles. The multivesicular bodies consist of dark, rounded structures containing several brilliant, globular microvesicles. Dark elongated vacuoles extending almost completely between the luminal surface and the basement membrane are also present.

The SEM fractographs of the capillary cross-section allowed us to study the trabecular, continuous, and interconnected system formed by rough endoplasmic reticulum of endothelial peripheral zone. The longitudinal fractured endoplasmic canaliculi exhibit a pierced or pitted surface, apparently the points of continuity with neighboring endoplasmic vacuoles and vesicles. Between the endoplasmic reticulum elements, the micropinocytotic vesicles appear as free, spherical, or long-chained beaded-shaped structures.

The SEM fractographs of the capillary lumen of *Salmo trutta* fixed by perfusion with Karnovsky's fixative and frozen with Freon 22 (Figure 85) show the points of capillary bifurcations. The longitudinal and transverse pathways of endothelial junctions appear as fine lines abutting the luminal surface. The pits or stomata display a protruding edge.

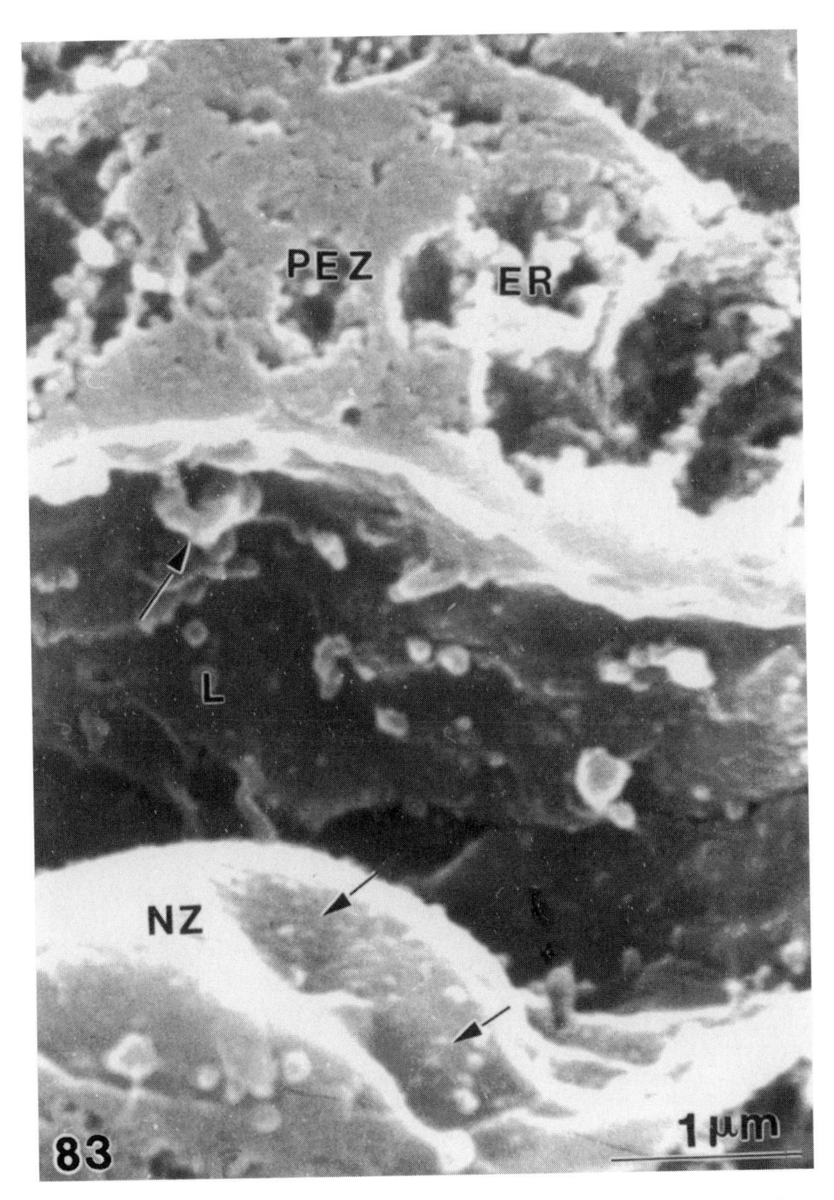

FIGURE 83. Liquid nitrogen frozen teleost fish *Arius spixii* cerebellar cortex. Scanning electron microscopy fractograph of endothelial nuclear zone (NZ) showing the luminal surface of deep invaginations (short arrows). The fractured peripheral endothelial zone (PEZ) containing endoplasmic reticulum (ER) profiles. The long arrow indicates a microvilli. Gold–palladium coating.

These features were clearly appreciated in the perfused material of *S. trutta*, where the perfusate fixative apparently "washes out" the endocapillary layer (Luft, 1971) and plasma substance and smoothes the endothelial luminal surface. In fortuitous SEM fractographs where the endothelial cells became detached by the fracture process, the

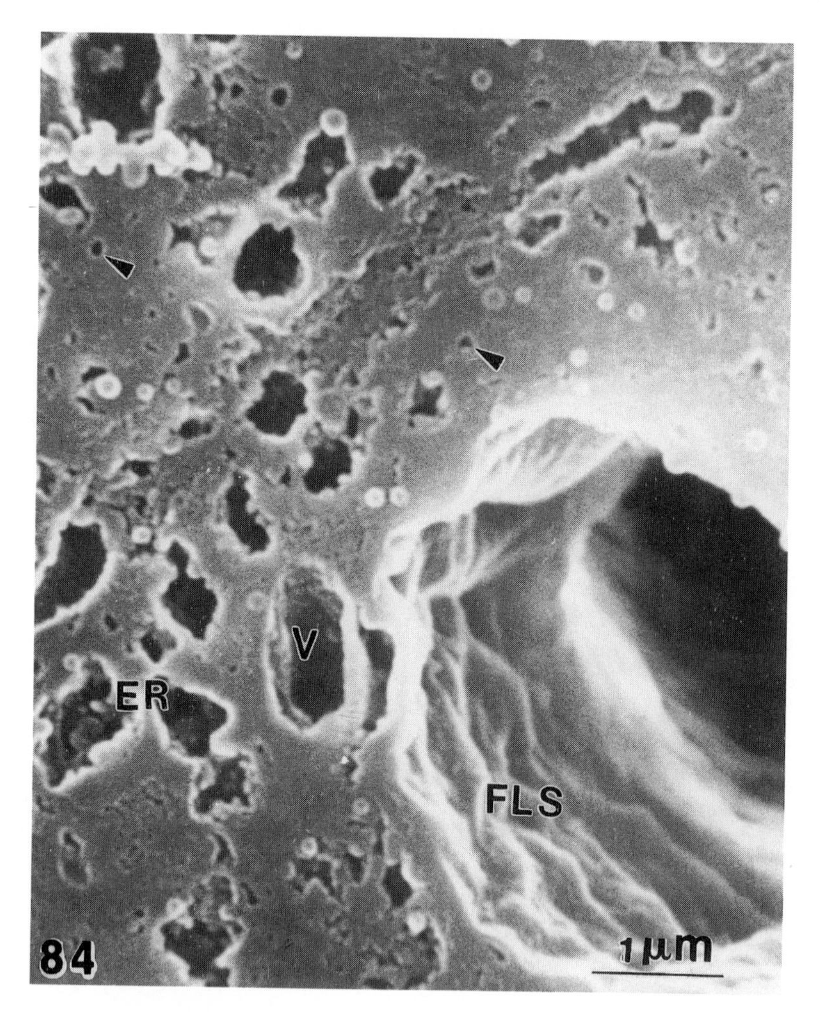

Figure 84. Liquid nitrogen frozen teleost fish *Arius spixii* cerebellar cortex. Scanning electron microscopy fractograph of capillary cross-section showing folded endothelial luminal surface (FLS), endoplasmic reticulum profiles (ER), vacuoles (V), and micropinocytotic vesicles (arrowheads). The endoplasmic reticulum canaliculi, cisternae, and vesicles form a trabecular system embedded in cytosol with a glassy appearance. Gold–palladium coating.

basement membrane inner surface exhibits a homogeneous, compact matrix interrupted by the enclosed pericytal processes (Figure 86). The basement matrix fibers occur as randomly orientated fine filaments.

The compact perivascular neuropil is formed by perivascular astrocytes and axonal and dendritic processes (Figure 87). In some fractographs devoid of perivascular neuropil, longitudinally and perpendicularly oriented astrocytic end-feet can be observed intimately applied to the capillary outer surface (Figure 88). At higher magnification, several astrocytic end-feet processes are observed apposed to the basement membrane in an overlapping configuration and entirely covering the outer capillary surface (Figure 89).

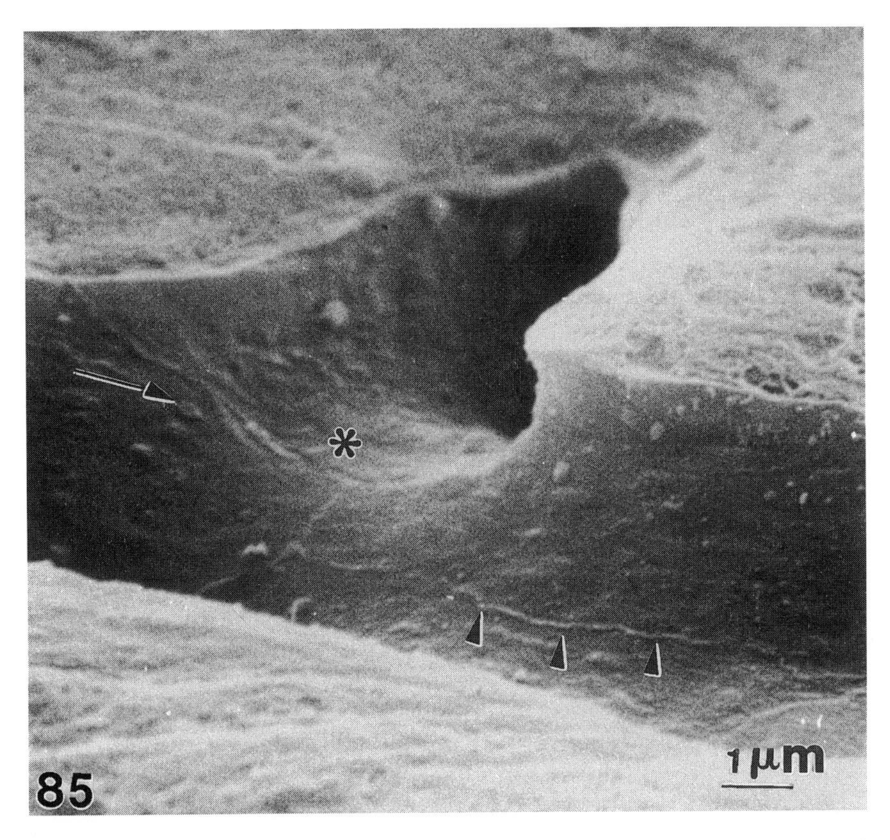

FIGURE 85. Freon frozen-fractured *Salmo trutta* cerebellar cortex. Scanning electron microscopy fractograph of the smooth endothelial surface of a capillary longitudinally fractured at its bifurcation point (asterisk). The endothelial borders appear as a protruding line (arrowheads). A prominent stoma, corresponding to an open micropinocytotic vesicle is observed (arrow). Gold–palladium coating.

CONTRIBUTION OF SEM TO THE CEREBELLAR BLOOD–BRAIN BARRIER STRUCTURE AND FUNCTION

SEM provides the three-dimensional morphological substrate for the transcapillary exchange or transcytosis process (Simionescu, 1980). As formerly observed by Simionescu et al. (1974) in muscle capillaries and subsequently confirmed by Castejón (1980) in human brain capillaries, the peripheral zone of endothelial cells plays a fundamental role in transendothelial transport whereas the nuclear zone has only a slight functional activity. Most endothelial vacuoles and micropinocytotic vesicles hitherto described in endothelial cells belong to the clear type (Brightman, 1967; Bruns & Palade, 1968; Casley-Smith, 1976; Donahue & Pappas, 1961; Westergaard, 1977; Westergaard & Brightman, 1973). Presumably, the dense vesicles are related to the presence of a highly electron dense plasma substance in teleost fishes, whereas the clear vesicles seem to transport proteins, water, ions, and small molecules. The highly electron dense plasma

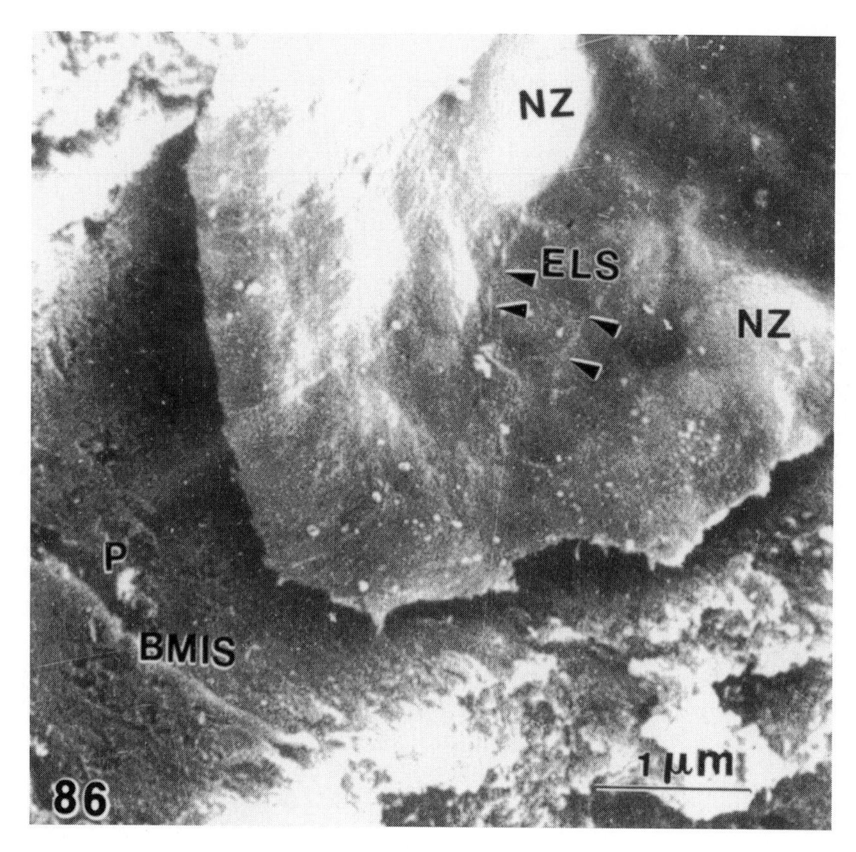

FIGURE 86. Freon frozen-fractured *Salmo trutta* cerebellar capillary. Scanning electron microscopy fractograph viewed from the capillary luminal surface. A segment of the endothelial luminal surface (ELS) shows the three-dimensional relief of the prominent nuclear endothelial zones (NZ). The freeze–fracture method removed the peripheral endothelial cytoplasm thus exposing the basement membrane inner surface (BMIS). The homogeneous texture of basement membrane matrix is observed. Enclosed pericyte processes (P) also are noted. Gold–palladium coating.

substance can be considered as an endogenous tracer that can be used for future capillary permeability studies in teleost fishes.

SEM fractographs of endothelial cells give the impression that, in a dynamic state, endothelial vacuoles and micropinocytotic vesicles form a continuous, anastomotic communication system with the rough endoplasmic structures, and that we are not dealing with isolated cytoplasmic organelles. Besides, some micropinocytotic vesicles and vacuoles are observed coalescing or fusing with the endoplasmic reticulum membranes.

The SEM appearance of interendothelial junctions, as depicted in Figure 85, supports the concept of Reese and Karnovsky (1967) that the endothelial junctions represent the morphological locus of blood–brain barrier, contrasting with the endothelial peripheral cytoplasm, which evidently represents the pore system (Pappenheimer, 1953) or active site of transcapillary exchange process. A clear visualization of the endothelial cell boundary line is obtained with SEM which closely resembles the description of artery or sinusoid

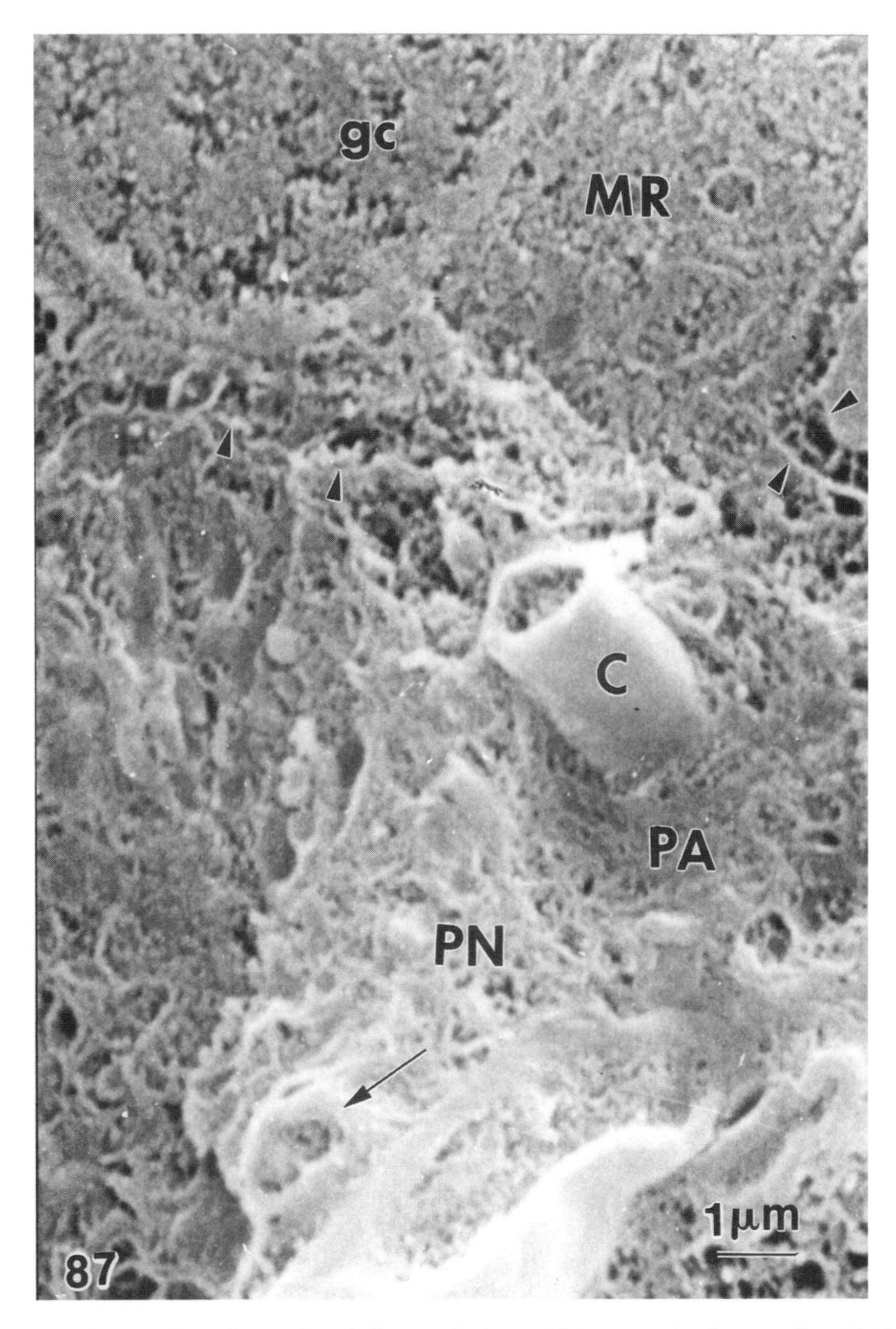

FIGURE 87. Freon frozen-fractured cerebellar granular layer of *Salmo trutta* showing a capillary (C). The compact perivascular neuropil (PN) is formed by perivascular astrocytic processes (PA), myelinated axons (arrow), and dendritic processes (arrows). A fractured granule cell (gc) and a mossy fiber rosette (MR) also are distinguished. Gold–palladium coating.

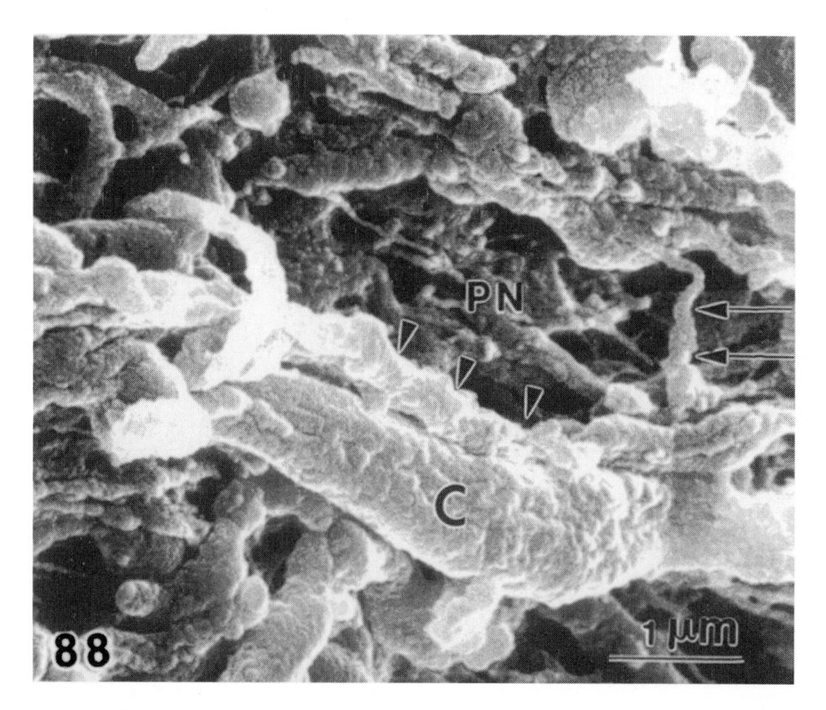

FIGURE 88. Liquid nitrogen frozen-fractured teleost fish *Arius spixii* cerebellar cortex showing the capillary (C) outer surface and the longitudinally applied perivascular astrocytic end-foot process (arrowheads). Another perpendicularly orientated astrocytic end-foot process (arrows) also is seen. A partial view of the perivascular neuropil (PN) is seen. Gold–palladium coating.

endothelial borders, as shown by Gertz, Rennels, Forbes, and Nelson (1975), Becker and De Bruyn (1975), and Albert and Nayak (1976).

In SEM fractographs, the endothelial cytoplasmic matrix or cytosol exhibits a smooth, glassy structure, closely resembling that observed by Humphreys et al. (1974) in rat hepatocytes. This cytosol appearance suggests that minimum artifacts have been introduced by the SEM preparative procedures.

The capillary luminal surface exhibits a different aspect in the samples processed by the immersion fixation technique than in those processed by vascular perfusion. In this latter method, the luminal surface exhibits a smooth, distended appearance, apparently induced by the pressure of the perfusate fixative. The former method induced a slightly folded appearance of the luminal surface. Gertz, Rennels, Forbes, and Nelson (1975) emphasized that in situ fixation by intravascular perfusion is superior to immersion, for the preservation of endothelial surface morphology. This also is the consensus according to most transmission electron microscopic (TEM) studies. The teleost fish *Arius spixii* displayed circumferential endothelial folds resembling the longitudinal folds formerly described by Kawamura, Gertz, Sunaga, Rennels, and Nelson (1974). Albert and Nayak (1976) suggested that such longitudinal folds were artifacts due to endothelial cell contraction and resulting from the preparative procedures. As in the present study, several investigators have found that these folds are significantly reduced or disappear by perfusion fixation (Clark & Glasgow, 1976; Swinehart, Pently, & Kardong, 1976; Thurston,

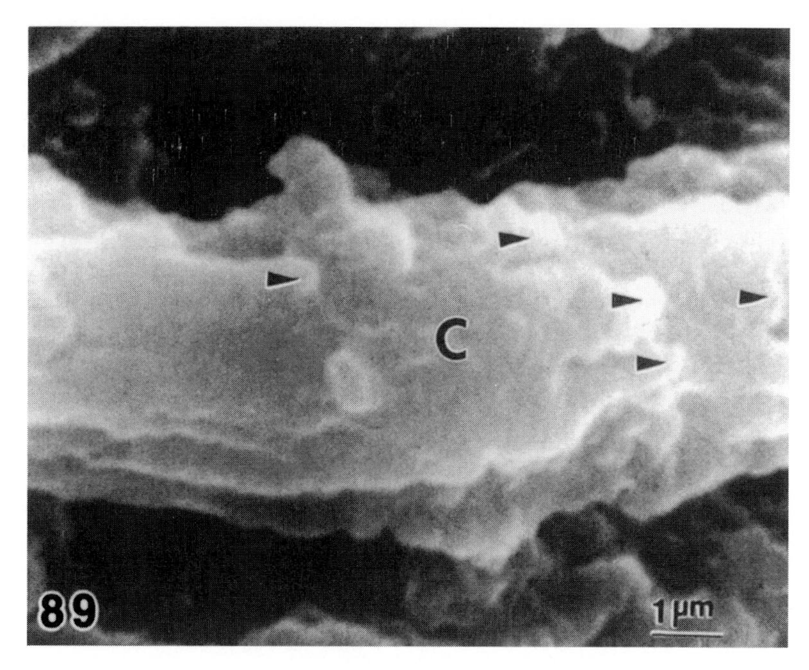

FIGURE 89. Liquid nitrogen frozen teleost fish *Arius spixii* cerebellar cortex. Scanning electron microscopy fractograph of a longitudinally oriented capillary (C) shows its outer surface devoid of perivascular neuropile. The astrocytic end-feet (arrowheads) are attached to its outer surface in an overlapping arrangement. Gold–palladium coating.

Buncke, Chater, & Weinstein, 1976). Lee and Chieng (1979) have demonstrated that the luminal surface undergoes a gradual flattening of the endothelial ridges with increasing transmural pressure. However, it remains to be demonstrated in cerebellar capillaries whether such folds are related to regional permeability as shown by Katora and Hollis (1976) in rat aorta.

In the present study, we have reported endothelial protrusions or microvilli which have also been previously described in normal and experimental endothelial surfaces of arteries (Fujimoto, Yamamato, & Takeshige, 1975; Smith, Ryan, Michic, & Smith, 1971; Still & Dennison, 1974). Other authors have suggested that these structures may be artifacts (Clark & Glasgow, 1976; Wolinsky, 1972). Our previous studies in human capillaries indicate that microvilli are mainly related to the formation of endothelial vacuoles (Castejón, 1980) especially in areas of increased cerebrovascular permeability. In normal conditions, endothelial microvilli are indicative of variations of regional permeability, for example, vacuolar transport. The SEM appearance of the basement matrix closely resembled that shown by Bauman et al. (1978) in dog renal artery and seems to play, along with the tight endothelial junctions, a primary role in blood–brain barrier phenomena.

The SEM appearance of the perivascular astrocytic end-feet completely ensheathing the capillary wall, as depicted in Figure 89, favors from the structural point of view the earlier hypothesis that the astrocytic end-feet play a role in the blood–brain barrier (Peters, Palay, & Webster, 1970).

CONCLUDING REMARKS

Using the freeze–fracture method, the SEM of cerebellar capillaries, reveals the luminal and inner surfaces of endothelial cells in cross and longitudinally fractured capillaries. Some structures responsible for the transcytosis process, such as endothelial vacuoles and micropinocytotic vesicles were identified. The three-dimensional relief of interendothelial junctions, the morphological locus of the blood–brain barrier, was observed in the luminal surface. The perivascular astrocytic end-feet overlap and are closely associated with the capillary basement membrane outer surface. The capillary structures are correlated with the blood–brain barrier phenomena.

Chapter 14

Contribution of Scanning Electron Microscopy to Cerebellar Neurobiology

INTRODUCTION

During the last three decades, progress in the development and application of biological techniques for scanning electron microscopy (SEM) has resulted in a marked increase in knowledge and comprehension of nerve tissue organization. An increasing number of investigations have attempted to reveal, the three-dimensional cellular SEM organization of different regions of the central nervous system (Abbate, Laura, Muglia, Vita, & Bronzetti, 1993; Allen & Didio, 1977; Antal, 1977; Asou, Murakami, Toda, & Uyemura, 1995; Baluk & Fujiwara, 1984; Berry et al., 1998; Bertmar, 1972; Boyde & Wood, 1969; Bredberg, 1979; Castejón, Castejón, Díaz, & Castellano, 2001d; Clementi & Marini, 1972; Cloyd & Low, 1974; Eagleson & Malacinski, 1986; Estable-Puig & Estable-Puig, 1975; Fasekas, Bacsy, & Rappay, 1978; Galbavy & Olson, 1979; Gross & De Boni, 1990; Hansson, 1970; Hartwig & Korf, 1978; Heavner, Coates, & Pacz, 2000; Kessel & Kardon, 1981; Kido & Sekitane, 1994; Kido, Sekitani, Okami, Endo, & Moriya, 1993; Kim, Kim, Moretto, & Kim, 1985; Krstic, 1974; Lametschwandtner, Simonsberg, & Adam, 1977; Lazzari & Franceschini, 2000; Lewis, 1971; Livesey & Fraher, 1992; Low, 1982; Mathew, 1998; Matsuda & Uehara, 1981; Meller & Denis, 1993; Merchant, 1979; Mestres & Rascher, 1994; Obsuki, 1972; Phillips, Balhorn, Leavitt, & Hoffman, 1974; Reina, Dittmann, López Garcia, & Van Zundert, 1997; Riesco et al., 1993; Saetersdal & Myklebust, 1975; Sanchez, Bilinski, González Nicolini, Villar, & Tramezzani, 1997; Sarphie, Carey, Davidson, & Soblosky, 1999; Scott, Kozlowski, Paul, Ramalingam, & Krobisch-Dudley, 1973; Seymour & Berry, 1975; Shotton, Heuser, Reese, & Reese, 1979; Siew, 1979; Tachibana, Takeuchi, & Fujiwara, 1985; Takahashi-Iwanaga, Murakami, & Abe, 1998; Takumida, Miyawaki, Harada, & Anniko, 1995; Tamega, Tirapelli, & Petroni, 2000; Tanaka, Iino, & Naguro, 1976; Yamadori, 1972). However, as the reader can appreciate throughout this monograph, very few papers have been devoted to the study of cerebellar cortex organization by conventional and high resolution SEM.

THE CHARACTERIZATION OF AFFERENT AND EFFERENT FIBERS IN THE CEREBELLAR WHITE MATTER

As described in Chapter 2, conventional SEM using slicing and cryofracture methods allows for identification of afferent mossy and climbing fibers and efferent Purkinje cell axons in the white matter. Identification of these fibers was made taking into account the Golgi LM criteria provided by published microscopical studies which highlight fiber thickness and branching pattern. Secondary electron images can be used for morphometric studies on cerebellar afferent fibers and the functional organization of these compartments (Feirabend, Choufoer, & Voogd, 1996).

THREE-DIMENSIONAL VISUALIZATION OF UNFRACTURED AND FRACTURED NEURONS

The slicing technique for conventional SEM demonstrated the outer surfaces of neurons and glial cells. Ethanol-cryofracturing and the freeze–fracture method displayed the spatial distribution of the endoplasmic reticulum, cell organelles, cytoskeletal structures, and nuclear chromatin organization of cerebellar neurons. In this context, SEM images could be compared those obtained with Golgi LM, high voltage EM (Palay & Chan-Palay, 1974), and confocal laser scanning microscopy (Castejón & Sims, 1999, 2000; Castejón & Castejón, 2000; Castejón, Castejón, & Sims, 2001a).

Three-dimensional visualization of unfractured and fractured neurons produced by various methods provides a new view of granule cell groups and their modular organization. This new insight explains how granule cell groups interact and communicate in the granular layer which, in turn, helps cerebellar scientists to construct a more detailed hypothesis concerning information processing in the granular layer. The SEM freeze–fracture method showed us the outer surface of Purkinje cells as well as the supra- and infraganglionic plexuses and a partial view of the Purkinje pericellular nest. In addition, it displayed the outer and inner surfaces of parallel fibers in the molecular layer and their synaptic relationships with Purkinje dendritic spines. The images of cross-fractured parallel fibers offer the possibility for accurately measuring the diameter of parallel fibers and shed light on their conduction velocity, a critical parameter for the theory of timing in the cerebellar cortex (Sultan, 2000).

The freeze–fracture method for SEM exposed the three-dimensional arrangement of the endoplasmic reticulum of cerebellar neurons, which forms a continuous compartment extending from the cell body toward the dendritic ramifications. This continuous compartment may coordinate and integrate neuronal function (Terasaki, Slater, Feiw, Schmidek, & Reese, 1994).

The SEM micrographs of basket and stellate cells in the deep molecular layer offer substantial proof that both cell types can be classified into distinct cell groups, a fact recently questioned by Sultan and Bower (1998).

SEM AS A HIGH RESOLUTION TOOL FOR TRACING CEREBELLAR INTRACORTICAL CIRCUITS

SEM microscopy and the used sample preparation methods allowed tracing the incoming mossy fibers in the granular layer and the climbing fibers in the three-layered structure of cerebellar cortex. The *en passant* nature of mossy fiber contacts with granule cell groups demonstrates a higher degree of divergent information of mossy fibers in the granular layer than that provided by classical light microscopy (LM) and transmission electron microscopy (TEM) studies. This supports recent analysis carried out with computer-assisted microscopy (Hámori, Jakab, & Takacs, 1997) and confocal laser scanning microscopy (Castejón & Sims, 1999, 2000; Castejón, Castejón, & Sims, 2000a; Castejón, Castejón, & Alvarado, 2000b; Castejón, Castejón, & Sims, 2001a). The SEM images provided basic morphological information that can be used for future studies on mossy fiber–granule cell synaptic relationship using image analysis and quantitative morphometric methods.

The SEM micrographs of climbing fiber pathway in the three-layered structure of cerebellar cortex show a sagittal zonal organization, and add further information on cerebellar zonal (Tan, Gerritz, Nanhoe, Simpson, & Woogd, 1995; Trott, Apps, & Armstrong, 1998), microzonal (Massion, 1993), sagittal (Garwicz, 1997), and parasagittal organization (Schweighofer, 1998).

An outstanding contribution of SEM is the finding of a high degree of lateral collateralization of climbing fibers in the granular and molecular layers, and this offers new views on the participation of these fibers on information processing in the cerebellar cortex (De Schutter & Maek, 1996). It allows for in-depth analysis of the synaptic relationships with granule, Golgi, Purkinje, basket, and stellate cells. These intracortical circuits deserve further investigations of new fractured hidden surfaces of cerebellar neurons using FESEM. Another conspicuous contribution of SEM was the tracing of intrinsic cortical circuits, such as basket cell–Purkinje cell synaptic contacts, which gives additional evidences for a unitary relationship between both cells (O'Donohue, King, & Bishop, 1989). The SEM examination of granule cell–Purkinje spine synapses and of basket cell axonal collaterals embracing the Purkinje cell soma contributing to the pericellular nest, reveals the potential contribution of SEM for tracing short intracortical circuits.

As certainly expressed by Llinás (1984) "probably the most striking example of uniformity in the neuronal fabric of the brain is that present in the cerebellar cortex. Its connectivity and neuronal circuitry have an almost crystal-like structural organization. The parallel fibers course in parallel to the cerebellar surface, the basket cells run orthogonally with respect to the direction of parallel fibers, and all dendritic processes: Purkinje dendritic branches, Golgi ascending dendrites, basket and stellate dendrites run radially toward the surface of the folium." In the present study, this beautiful geometrical arrangement facilitated their identification by means of SEM.

THE THREE-DIMENSIONAL MORPHOLOGY OF SYNAPTIC CONNECTIONS

The cryofracture method for conventional SEM made it possible to image the outer and inner surfaces of parallel fibers in the molecular layer and their synaptic relationships with the Purkinje dendritic spines. FESEM yielded SE-I images, which exhibited the three-dimensional relief of synaptic vesicles contained in the parallel fiber varicosities. High magnification FESEM resolved the structure of the active synaptic membrane complex formed by pre- and postsynaptic membranes, the synaptic cleft and the globular subunits present at the postsynaptic density. These SE-I images for the first time showed the three-dimensional configuration of a synaptic membrane complex, in agreement with previous

TEM descriptions made with ultrathin sections or freeze-etching direct replicas. Such findings open new lines of microscopical research on the contribution of SEM cerebellar synaptic morphology to cognitive processes, involving motor learning and long-term memory (Anderson, Alcántara, & Greenough, 1996; De Zeeuw, Simpson, Hoogenraad, Galjart, Koekkoek, & Ruigrok, 1998; Hansen & Linden, 2000; Hesslow, Svensson, & Ivarsson, 1999; Kenyon, 1997; Kleim, Swain, Armstrong, Napper, Jones, & Greenough, 1998).

THE THREE-DIMENSIONAL MORPHOLOGY OF GLIAL CELLS

Scanning electron micrographs showed the outer and inner surface features of Bergmann cells, oligodendrocytes, and velate protoplasmic astrocytes. The topographic relationship of Bergmann cells with Purkinje cells provided the morphological correlate of neuron-glial unit in the Purkinje cell and molecular layers.

CONTRIBUTION TO THE INFORMATION PROCESSING IN THE CEREBELLAR CORTEX

The anatomy of the cerebellar cortex can be described as a two-layered network. The input layer, corresponding to the granular layer, processes the incoming mossy fiber signal and transmits them by the parallel fiber system to the output layer, which consists mainly of Purkinje cells. In both layers, activity is controlled by inhibitory neurons, the Golgi cells, in the input layer. Mossy fibers activate both the excitatory granule cells and the inhibitory Golgi cells. The granule cell axons form the parallel fibers, which not only transmit information to the output layer, but also provide additional excitatory input to Golgi cells (De Schutter, Vos, & Maex, 2000). SEM and cryofracture techniques have offered "*en face*" and saggital views of mossy fibers and their synaptic relationship with up to 18 granule cells. These images on the degree of divergence of excitatory information of mossy fibers to granule cells is closely similar to that reported by Eccles, Ito, and Szentágothai (1967) of 15 granule cells entering a single glomerulus. As demonstrated earlier, by classical electrophysiological studies of these authors, the mossy fibers excite granule and Golgi cells in the granular layer. Climbing fibers establish synaptic contacts and exert strong excitatory action on the Purkinje cell primary dendritic trunk and secondary dendritic spiny branchlets (Eccles, Ito, & Szentágothai, 1967; Ito, 1984). This mode of functioning would be expected to increase cerebellar learning capacity (Ekerot, 1985). The climbing fibers make excitatory connections to all the cerebellar inhibitory neurons: Golgi, Lugaro, Purkinje, stellate, and basket cells. Golgi cells, in turn, discharge inhibition on excitatory granule cells and in the mossy glomerular regions.

We have also demonstrated by conventional SEM the 1 to 1 synaptic relationships between Golgi cells and granule cells in the granule layer. Golgi cells that are activated by the parallel fiber in the molecular layer, inhibit granule cells and prevent their activation by mossy fibers (Cohen & Yarom, 2000). As mentioned above, SEM reveals a high degree of lateral collateralization of climbing fibers in the granular and molecular layers. The climbing fiber input to the cerebellum imposses a parasagittal organization on the cerebellar cortex (Groenwegen & Voogd, 1977; Voogd & Glickstein, 1998). This organization has been also seen with a number of staining techniques (Hawkes, Brochin, Dore, Gravel, & Leclere,

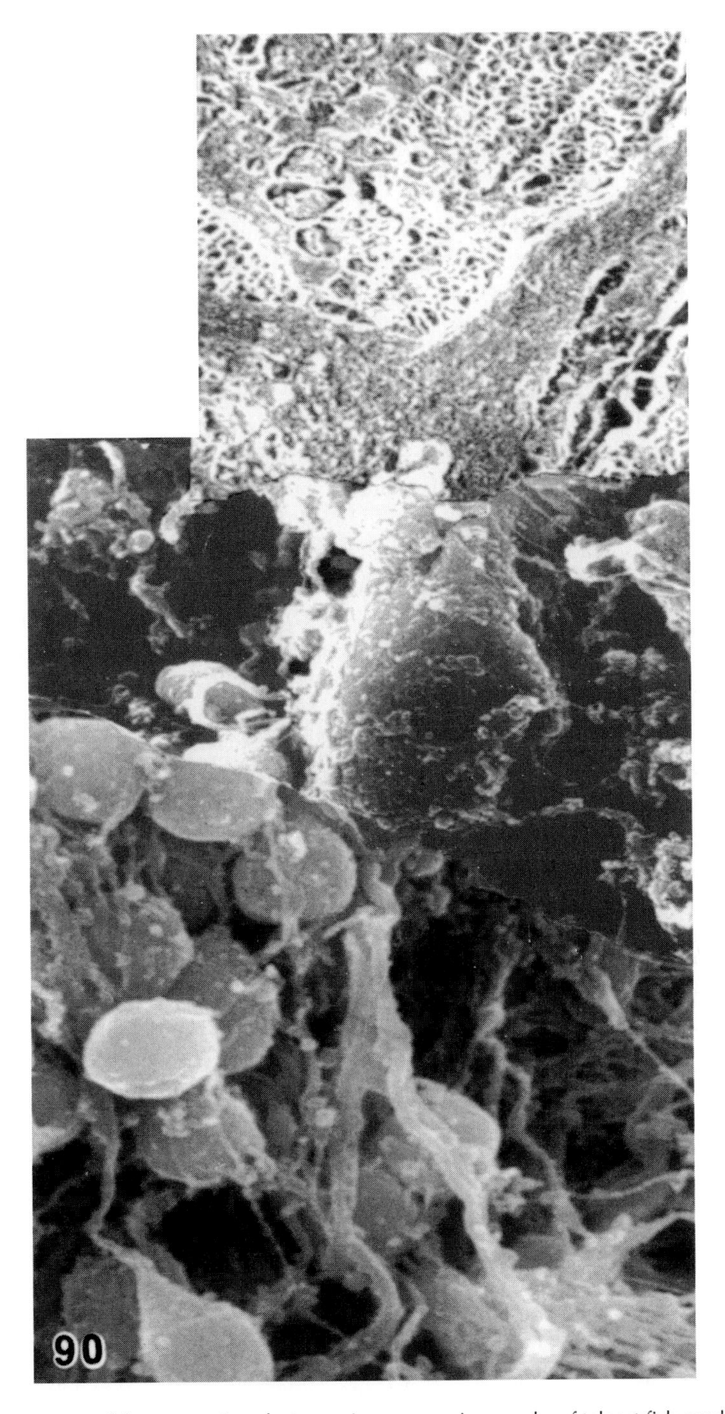

FIGURE 90. Montage of three scanning electron microscopy micrographs of teleost fish cerebellar cortex to illustrate a model of the three-dimensional structure of the three-layered structure of cerebellar cortex. Gold–palladium coating.

1992; Shinoda, Sughihara, Wu, & Sugiuchi, 2000), and is, in addition, supported by our SEM findings (Castejón et al., 2000a, 2000b, 2001a; Castejón, Apkarian, Castejón, & Alvarado, 2001b). The morphological sagittal organization also corresponds with a functional organization (Llinás & Sasaki, 1989).

The three-dimensional representation of granule cells, as depicted in Chapter 3, favors the microzone concept (Ito, 1984; Oscarsson, 1979) and is in agreement with the mode of function of corticonuclear microcomplexes, which would represent a modular structure of the cerebellum. A modular organization of the cerebellum has been postulated by several investigators (Buisseret-Delmas & Angaut, 1993; Hawkes & Gravel, 1991; Oscarsson, 1969; Teune, Van Der Burg, Van Der Moer, Woogd, & Ruigrok, 2000; Tolbert, 1978; Voogd, 1967). Each module is formed by a localized group of granule cells and the Purkinje cells above them. Each module is activated by two inputs: the mossy fibers and climbing fibers. These modules perform a computational process that may participate in the various functions attributed to the cerebellum (Mann-Metzer & Yarom, 2000). Our SEM micrographs give support to the zonal hypothesis or modular organization of the cerebellar cortex.

Basket and stellate cells inhibit Purkinje cells, which represent the only output of cerebellar cortex, that exerts inhibition on deep cerebellar nuclei. The reader is addressed to the Castejón et al. papers (2000a, 2000b; Castejón, Castejón & Apkarian, 2001c), for more elaborate neurobiological considerations on mossy and climbing fibers and information processing in the cerebellar cortex.

THE THREE-DIMENSIONAL DESIGN OF THE CEREBELLAR CORTEX

As illustrated in Figure 90, which represents a composite mosaic of SEM micrographs of teleost fish cerebellar cortex, an almost complete SEM three-dimensional image of the cytoarchitectonic arrangement of the cerebellar cortex can be obtained. Such a new view has seminal value in neuroanatomy and neurohistology to construct a three-dimensional model of the cerebellar cortex, which can be used for theoreticians dealing with information processing in the cerebellum. If this goal is achieved, the work involved in obtaining a SEM three-dimensional view of cerebellar cortex was well-warranted.

References

Abbate, F., Laura, R., Muglia, U., Vita, G., & Bronzetti, P. (1993). Differentiation of ependymal surface of lateral ventricles in fetus and newborn rabbits: Observations by SEM. *Anatomía, Histología, Embriología, 22*, 348–354.

Abbott, L. C., & Sotelo, C. (2000). Ultrastructural analysis of catecholaminergic innervation in weaver and normal mouse cerebellar cortices. *Journal of Comparative Neurology, 426*, 316–329.

Akima, M., Nonaka, H., Kagesewa, M., & Tanaka, K. (1987). A study on the microvasculature of the cerebellar cortex. The fundamental architecture and its senile change in the cerebellar hemisphere. *Acta Neuropathologica, 75*, 69–76.

Albert, E. N., & Nayak, R. K. (1976). Surface morphology of human aorta as revealed by the scanning electron microscope. *Anatomical Record, 185*, 223–234.

Allen, J. A., & Didio, L. J. A. (1977). Scanning and transmission electron microscopy of the encephalic meninges in dogs. *Journal of Submicroscopic Cytology and Pathology, 9*, 1–22.

Alvarez-Otero, R., & Anadon, R. (1992). Golgi cells of the cerebellum of the dogfish, *Scyliorhinus Canicula* (elasmobranch): A Golgi and ultrastructural study. *Journal für Hirnforschung, 33*, 321–327.

Anderson, T. F. (1951). Techniques for the preservation of three-dimensional structures in preparing specimens for the electron microscope. *Transaction of the New York Academy of Sciences, 13*, 130–134.

Anderson, B. J., Alcántara, A., & Greenough, W. T. (1996). Motor-skill learning changes in synaptic organization of the rat cerebellar cortex. *Neurobiology of Learning and Memory, 66*, 221–229.

Antal, M. (1977). Scanning electron microscopy of photoreceptors. *Ophtalmologica (Basel), 174*, 280–284.

Apergis, G., Alexopoulos, T., Mpratokos, M., & Katsorchis, T. (1991). Scanning electron microscopy of the granular layer of rat cerebellar cortex. *Microscopía. Electrónica y Biología Celular, 15*, 119–129.

Apkarian, R. P., & Curtis, J. C. (1986). Hormonal regulation of capillary fenestrae in the rat adrenal cortex: Quantitative studies using objective lens staging scanning electron microscopy. *Scanning Electron Microscopy, 4*, 1381–1393.

Apkarian, R. P., & Joy, D. C. (1988). Analysis of metal films suitable for high-resolution SE-I microscopy. In D. Newbury (Ed.), *Microbeam analysis* (pp. 459–462). San Francisco: San Francisco Press.

Arnett, C. E., & Low, F. N. (1985). Ultrasonic microdissection of rat cerebellum for scanning electron microscopy. *Scanning Electron Microscopy, 1*, 247–255.

Arro, E., Collins, V. P., & Brunk, U. T. (1981). High resolution SEM of cultured cells: Preparatory procedures. *Scanning Electron Microscopy, 2*, 159–168.

Asou, H., Murakami, K., Toda, M., & Uyemura, K. (1995). Development of oligodendrocyte in the central nervous system. *Keio Journal of Medicine, 44*, 47–52.

Athias, M. (1897). Recherches sur l'histogénèse de l'écorce du cervelet. *Journal de Anatomie et Physiologie Normal et Pathologica de l'Homme et des Animaux, 33*, 372–399.

Baluk, P., & Fujiwara, T. (1984). Direct visualization by scanning electron microscopy of the preganglionic innervation and synapses on the true surfaces of neurons in the frog heart. *Neuroscience Letter, 51*, 265–270.

Barber, V. C., & Graziadei, P. (1967). The structure of cephalopod blood vessels. II. The vessels of the nervous system. *Zeitschrift fur Zellforschung, 77*, 147–161.

Bauman, F. G., Imparato, A. M., Kim, G. E., Yoder, M., & Grover-Thonson, M. (1978). A study of evolution of early fibromuscular lesions hemodynamically induced in the dog renal artery. II. Scanning and correlative transmission electron microscopy. *Artery, 4*, 67–99.

Becker, R. P., & De Bruyn, P. P. H. (1975). The transmural passage of blood cells into myeloid sinusoids and the entry of platelets into the sinusoidal circulation. A scanning electron microscopic investigation. *American Journal of Anatomy, 145*, 183–206.

Bell, P. B. (1984). The preparation of whole cells for electron microscopy. In J. P. Revel, T. Bernard, & G. H. Haggis (Eds.), *The science of biological specimen preparation for microscopy and microanalysis* (pp. 45–59). Scanning Electron Microscopy. Chicago: Scanning Microscopy International.

Bergmann, C. (1857). Notiz über einige Structurverhältnisse des Cerebellum und Rückenmarks. *Zeitschrift für Rationelle Medizin* (Henle und Pfeufer), N.F.8, 360–363.

Berliner, K. (1905). Beiträge zur Histologie und Entwickelungsgeschichte des Kleinhirns. *Archives für mikroskopische Anatomie, 66*, 220–269.

Berry, M., Hunter, A. S., Duncan, A., Lordan, J., Kirvell, S., Tsang, W. L., & Butt, A. M. (1998). Axon-glial relations during regeneration of axons in the adult rat anterior medullary velum. *Journal of Neurocytology, 7*, 915–937.

Bertmar, G. (1972). Scanning electron microscopy of olfatory rosette in sea trout. *Zeitschrift für Zellforschung, 128*, 336–346.

Bielschowsky, M., & Wolff, M. (1904). Zur Histologie der Kleinhirnrinde. *Journal of Psychology, 4*, 1–23.

Bishop, G. A. (1993). An analysis of HRP-filled basket cell axons in the cat's cerebellum. I. Morphometry and configuration. *Anatomy and Embryology (Berlin), 188*, 287–297.

Black, J. T. (1974). The scanning electron microscope. Operating principles. In M. A. Hayat (Ed.), *Principles and techniques of scanning electron microscopy. Biological applications. Vol. I* (pp. 1–42). New York: Van Nostrand/Reinhold.

Boecke, J. (1942). Sur les synapses a distance. Les glomérules cérébelleuses, leur structure et leur development. *Schweizer. Archiv für Neurologie und Psychiatrie, 49*, 9–32.

Bovero, J. A., Ascher, J., Arregui, A., Rovainen, C., & Woolsey, T. A. (1999). Increased brain capillaries in chronic hypoxia. *Journal of Applied Physiology, 86*, 1211–1219.

Boyde, A. (1972). Biological specimen preparation in the scanning electron microscope: An overview. *Scanning Electron Microscopy, 1*, 257–264.

Boyde, A. (1978). Pros and cons of critical point drying and freeze drying for SEM. *Scanning Electron Microscopy, 2*, 303–314.

Boyde, A., & Maconnachie, E. (1981). Morphological correlations with dimensional change during SEM specimen preparation. *Scanning Electron Microscopy, 4*, 27–34.

Boyde, A., & Wood, C. (1969). Preparation of animal tissue for surface scanning electron microscopy. *Journal of Microscopy, 90*, 221–249.

Boyde, A., Bailey, E., Jones, E. J., & Tamarind, A. (1977). Dimensional changes during specimen preparation for scanning electron microscope. *Scanning Electron Microscopy, 1*, 507–518.

Bredberg, G. (1979). Scanning electron microscopy of nerves within Organ of Corti. *Otology, Rhinology and Laryngology, 217*, 321–330.

Brightman, M. W. (1967). Intracerebral movement of proteins injected into the blood and cerebrospinal fluid. *Anatomycal Record (Abstract), 157*, 219.

Brodal, A. (1969). *Neurological anatomy in relation to clinical medicine*. New York: Oxford University Press.

Brodal, J., & Drablos, P. A. (1963). Two types of mossy fiber terminals in the cerebellum and their regional distribution. *Journal Comparative Neurology, 121*, 173–187.

Brunk, U., Bell, P., Colling, P., Forsby, N., & Fredriksson, B. A. (1975). SEM of *in vitro* cultivated cells: Osmotic effects during fixation. *Scanning Electron Microscopy, 1*, 379–385.

Bruns, R. R., & Palade, G. E. (1968). Studies on blood capillaries. I. General organization of blood capillaries in muscle. *Journal of Cell Biology, 37*, 244–276.

Buisseret-Delmas, C., & Angaut, P. (1993). The cerebellar olivo-cortico-nuclear connections in the rat. *Progress in Neurobiology, 40*, 63–87.

Buss, H., Klose, J. P., & Hollweg, H. G. (1976). Endothelial surfaces of renal, coronary and cerebral artery. *Scanning Electron Microscopy, 2*, 217–233.

Carrea, R. M. E., Reissig, H., & Mettler, F. A. (1947). The climbing fibers of the simian and feline cerebellum: Experimental inquiry into their origin by lesions of the inferior olives and deep cerebellar nuclei. *Journal of Comparative Neurology, 87*, 321–365.

Casley-Smith, J. R. (1976). The functioning and interrelationships of blood capillaries and lymphatics. *Experientia, 32*, 1–12.

Castejón, O. J. (1968). Electron microscopic observations at the level of the molecular layer of the cerebellar cortex. *Investigación Clínica, 27*, 57–108.

Castejón, O. J. (1969). Ultrastructure of granule cell layer of human cerebellar cortex. Organization of granule cells. *Investigación Clínica, 29*, 29–46.

Castejón, O. J. (1971). Ultrastructure of human cerebellar glomeruli. *Investigación Clínica, 38*, 49–72.

Castejón, O. J. (1976). Ultrastructure of Golgi cells of the cerebellar cortex. *Boletín de la Academia de Ciencias Físicas, Matemáticas y Naturales, 107*, 67–110.

Castejón, O. J. (1980). Electron microscopic study of capillary wall in human cerebral edema. *Journal of Neuropathology and Experimental Neurology, 39*, 296–327.

Castejón, O. J. (1981). Light microscope, SEM and TEM study of fish cerebellar granule cells. *Scanning Electron Microscopy, 4*, 105–113.

Castejón, O. J. (1983a). Light, scanning and transmission electron microscopy study of fish cerebellar capillaries. *Scanning Electron Microscopy, 1*, 151–160.

Castejón, O. J. (1983b). Scanning electron microscope recognition of intracortical climbing fiber pathways in the cerebellar cortex. *Scanning Electron Microscopy, 3*, 1427–1434.

Castejón, O. J. (1984). Low resolution scanning electron microscopy of cerebellar neurons and neuroglial cells of the granular layer. *Scanning Electron Microscopy, 3*, 1391–1400.

Castejón, O. J. (1986). Freeze-fracture of fish and mouse cerebellar climbing fibers. A SEM and TEM study. In T. Imura, S. T. Maruse, & S. Susuki (Eds.), *Electron microscopy. Vol. IV* (pp. 3165–3166). Tokyo: Japanese Society of Electron Microscopy.

Castejón, O. J. (1988). Scanning electron microscopy of vertebrate cerebellar cortex. *Scanning Microscopy, 2*, 569–597.

Castejón, O. J. (1990a). Freeze-fracture scanning electron microscopy and comparative freeze-etching study of parallel fiber-Purkinje spine synapses of vertebrate cerebellar cortex. *Journal Submicroscopic Cytology and Pathology, 22*, 281–295.

Castejón, O. J. (1990b). Surface and membrane morphology of Bergmann glial cells and their topographic relationships in the cerebellar molecular layer. *Journal of Submicroscopic Cytology and Pathology, 22*, 123–134.

Castejón, O. J. (1990c). Scanning electron microscopy study of cerebellar synaptic junctions. In L. D. Peachey & D. B. Williams (Eds.), *Electron microscopy 1990* (pp. 148–149). San Francisco: San Francisco Press.

Castejón, O. J. (1991). Three-dimensional morphological analysis of nerve cells by scanning electron microscopy. A review. *Scanning Microscopy, 5*, 461–476.

Castejón, O. J. (1993). Sample preparation techniques for conventional and high resolution scanning electron microscopy of the central nervous system. The cerebellum as a model. *Scanning Microscopy, 7*, 725–740.

Castejón, O. J. (1996). Contribution of conventional and high resolution scanning electron microscopy and cryofracture technique to the study of cerebellar synaptic junctions. *Scanning Microscopy, 10*, 177–186.

Castejón, O. J., & Apkarian, R. P. (1992). Conventional and high resolution scanning electron microscopy of outer and inner surface features of cerebellar nerve cells. *Journal of Submicroscopic Cytology and Pathology, 24*, 549–562.

Castejón, O. J., & Apkarian, R. P. (1993). Conventional and high resolution field emission scanning electron microscopy of vertebrate cerebellar parallel fiber-Purkinje spine synapses. *Cellular and Molecular Biology, 39*, 863–873.

Castejón, O. J., & Caraballo, A. J. (1980a). Application of cryofracture and SEM to the study of human cerebellar cortex. *Scanning Electron Microscopy, 4*, 197–207.

Castejón, O. J., & Caraballo, A. J. (1980b). Light and scanning electron microscopic study of cerebellar cortex of teleost fishes. *Cell and Tissue Research, 207*, 211–226.

Castejón, O. J., & Castejón, H. V. (1972). Light microscope, cytochemistry and ultrastructural study of mouse cerebellar Golgi cells. *Revista de Microscopía Electrónica, 1*, 162–163.

Castejón, O. J., & Castejón, H. V. (1981). Transmission and scanning electron microscopy and ultracytochemistry of vertebrate and human cerebellar cortex. In S. Federoff (Ed.), *Glial and neuronal cell biology* (pp. 249–258). New York: Alan R. Liss.

Castejón, O. J., & Castejón, H. V. (1987). Electron microscopy and glycosaminoglycans histochemistry of cerebellar stellate neurons. *Scanning Microscopy, 1*, 681–693.

Castejón, O. J., & Castejón, H. V. (1988). Scanning electron microscopy, freeze-etching and glycosaminoglycan cytochemical studies of the cerebellar climbing fiber system. *Scanning Microscopy, 2*, 2181–2193.

Castejón, O. J., & Castejón, H. V. (1991). Three-dimensional morphology of cerebellar protoplasmic islands and proteoglycan content of mossy fiber glomerulus: A scanning and transmission electron microscopic study. *Scanning Microscopy, 5*, 477–494.

Castejón, O. J., & Castejón, H. V. (1997). Conventional and high resolution scanning electron microscopy of cerebellar Purkinje cells. *Biocell, 21*, 149–159.

Castejón, O. J., & Castejón, H. V. (2000). Correlative microscopy of cerebellar Golgi cells. *Biocell, 24*, 13–30.

Castejón, O. J., & Sims, P. (1999). Confocal laser scanning microscopy of hamster cerebellum using FM4-64 as an intracellular staining. *Scanning, 21*, 15–21.

Castejón, O. J., & Sims, P. (2000). Three-dimensional morphology of cerebellar climbing fibers. A study by means of confocal laser scanning microscopy and scanning electron microscopy. *Scanning, 22*, 211–217

Castejón, O. J., & Valero, C. J. (1980). Scanning electron microscopy of human cerebellar cortex. *Cell and Tissue Research, 212*, 263–374.

Castejón, O. J., Castejón, H. V., & Apkarian, R. P. (1994a). Proteoglycan ultracytochemistry and conventional and high resolution scanning electron microscopy of vertebrate cerebellar parallel fiber presynaptic endings. *Cellular and Molecular Biology, 40*, 795–801.

Castejón, O. J., Apkarian, R. P., & Valero, C. (1994b). Conventional and high resolution scanning electron microscopy and cryofracture techniques as tool for tracing short intracortical circuits. *Scanning Microscopy, 8*, 315–324.

Castejón, O. J., & Castejón, H. V., & Apkarian, R. P. (1994c). High resolution scanning electron microscopy features of primate cerebellar cortex. *Cellular and Molecular Biology, 40*, 1173–1181.

Castejón, O. J., Castejón, H. V., & Sims, P. (2000a). Confocal, scanning and transmission electron microscopic study of cerebellar mossy fiber glomeruli. *Journal of Submicroscopic Cytology and Pathology, 32*, 247–260.

Castejón, O. J., Castejón, H. V., & Alvarado, M. V. (2000b). Further observations on cerebellar climbing fibers. A study by means of light microscopy, confocal laser scanning microscopy and scanning and transmission electron microscopy. *Biocell, 24*, 197–212.

Castejón, O. J., Castejón, H. V., & Sims, P. (2001a). Light microscopy, confocal laser scanning microscopy, scanning and transmisión electron microscopy of cerebellar basket cells. *Journal of Submicroscopic Cytology and Pathology, 33*, 23–32.

Castejón, O. J., Apkarian, R. P., Castejón, H. V., & Alvarado, M. V. (2001b). Field emission scanning electron microscopy and freeze-fracture transmission electron microscopy of mouse cerebellar synaptic contacts. *Journal of Submicroscopic Cytology and Pathology, 33*, 289–300.

Castejón, O. J., Castejón, H. V., & Apkarian, R. P. (2001c). Confocal laser scanning and scanning and transmission electron microscopy of vertebrate cerebellar granule cells. *Biocell, 25*, 235–255.

Castejón, O. J., Castejón, H. V., Díaz, M., & Castellano, A. (2001d). Consecutive light microscopy, scanning-transmission electron microscopy and transmission electron microscopy of traumatic human brain oedema and ischaemic brain damage. *Histology and Histopathology, 16*, 1117–1134.

Chan-Palay, V., & Palay, S. L. (1970). Interrelations of basket cell axons and climbing fibers in the cerebellar cortex of the rat. *Zeitschrift Anatomische Entwicklungsgesch, 132*, 191–227.

Chan-Palay, V., & Palay, S. L. (1971a). Tendril and glomerular collaterals of climbing fibers in the granular layer of the rat's cerebellar cortex. *Zeitschrift Anatomische Entwicklungsgesch, 133*, 247–273.

Chan-Palay, V., & Palay, S. L. (1971b). The synapse "en marron" between Golgi II neurons and mossy fiber in the rat's cerebellar cortex. *Zeitschrift Anatomische Entwicklungsgesch, 133*, 274–287.

Chan-Palay, V., & Palay, S. L. (1972a). The stellate cells of the rat's cerebellar cortex. *Zeitschrift Anatomische Entwicklungsgesch, 136*, 224–240.

Chan-Palay, V., & Palay, S. L. (1972b). The form of velate astrocytes in the cerebellar cortex of monkey and rat: High voltage electron microscopy of rapid Golgi preparations. *Zeitschrift Anatomische Entwicklungsgesch, 138*, 1–19.

Chen, S., & Hillman, D. E. (1999). Dying-back of Purkinje cell dendrites with synapse loss in aging rats. *Journal of Neurocytology, 28*, 187–196.

Cho, Y., & De Bruyn, P. P. H. (1979). The endothelial structure of the postcapillary venules of the lymph node and the passage of lymphocytes across the venule wall. *Journal of Ultrastructural Research, 69*, 13–21.

Christensen, B. C., & Garbasch, C. (1972). A scanning electron microscopic study on the endothelium of the normal rabbit aorta. *Angiologica, 9*, 15–26.

Clark, E., & Glasgow, S. (1976). Vascular endothelium by scanning electron microscopy. Elimination of configurational artifacts. *British Journal of Experimental Pathology, 57*, 129–135.

Clementi, F., & Marini, D. (1972). The surface fine structure of the walls of cerebral ventricles and of choroid plexus in cat. *Zeitschrift für Zellforschung, 173*, 82–95.

Cloyd, M. W., & Low, F. N. (1974). Scanning electron microscopy of the subarachnoid space in the dog. I. Spinal cord levels. *Journal of Comparative Neurology, 153*, 325–368.

Cohen, A. L. (1974). Critical point drying. In M. A. Hayat (Ed.), *Principles and techniques of scanning electron microscopy. Biological applications. Vol. 1* (pp. 44–112). New York: Van Nostrand/Reinhold.

Cohen, A. L. (1977). A critical look at critical point drying theory, practice and artefacts. *Scanning Electron Microscopy, 1*, 525–536.

Cohen, A. L. (1979). Critical point drying. Principles and procedures. *Scanning Electron Microscopy, 1*, 525–536.

Cohen, D., & Yarom, Y. (2000). Unravelling cerebellar circuitry: An optical imaging study. In N. M. Gerrits, T. J. H. Ruigrok, & C. I. De Zeeuw (Eds.), *Cerebellar modules: Molecules, morphology and function* (pp. 107–114). Progress in Brain Research. Vol. 124. New York: Elsevier.

Cragie, E. H. (1926). Notes on the morphology of the mossy fibers in some birds and mammals. *Trabajos de los Laboratorios de Investigaciones Biológicas. (Madrid), 24*, 319–331.

Crevatin, J. (1898). Über die Zellen von Fusari und Ponti im Kleinhirn von Säugetieren. *Anatomischer Anzeiger*, Bd. XIV.

Dahl, V., Olsen, S., & Birch-Anderson, A. (1962). The fine structure of the granular layer in the human cerebellar cortex. *Acta Neurologica Scandinavica, 38*, 83–97.

Davies, P. F., & Bowyer, D. E. (1975). Scanning electron microscopy: Arterial endothelial integrity after fixation at physiological pressure. *Atherosclerosis, 21*, 465–469.

Del Cerro, M. (1974). Uptake of tracer proteins in the developing cerebellum, particularly by the growth cones and blood vessels. *Journal of Comparative Neurology, 157*, 245–279.

Denissenko, G. (1877). Zur frage über den Bau der Kleinhirnrinde bei en verschiedenen Klassen von Wirbeltieren. *Archiv für Mikroskopische Anatomie, 14*, 203–242.

De Schutter, E., & Maex, R. (1996). The cerebellum: Cortical processing and theory. *Current Opinion in Neurobiology, 6*, 759–764.

De Schutter, E., Vos, B., & Maex, R. (2000). The function of cerebellar Golgi cell revisited. In N. M. Gerrits, T. J. H. Ruigrok, & C. I. De Zeeuw (Eds.), *Cerebellar modules: Molecules, morphology and function* (pp. 81–93). Progress in Brain Research. Vol. 124. New York: Elsevier.

De Zeeuw, C. I., Simpson, J. I., Hoogenraad, C. C., Galjart, N., Koekkoek, S. K., & Ruigrok, T. J. (1998). Microcircuitry and function of the inferior olive. *Trends in Neuroscience, 21*, 391–400.

Dogiel, A. S. (1896). Die Nervenelemente in Kleinhirnrinde der Vögel und Säugethiere. *Archiv für Mikroskopische Anatomie, 47*, 707–719.

Donahue, S., & Pappas, G. D. (1961). The fine structure of capillaries in the cerebral cortex of the rat at various stages of development. *American Journal of Anatomy, 108*, 331–347.

Duvernoy, H., Deon, S., & Vannson, J. L. (1983). The vascularization of the human cerebellar cortex. *Brain Research Bulletin, 11*, 419–480.

Eagleson, G. W., & Malacinski, G. M. (1986). A scanning electron microscopy and histological study of the effects of the mutant eyeless (e/e) gene upon the hypothalamus in the Mexican axolotl *Ambystoma Mexicanum Shaw. Anatomical Record, 215*, 317–327.

Eccles, J., Ito, M., & Szentágothai, J. (1967). *The cerebellum as a neuronal machine* (pp. 4–226). New York: Springer-Verlag.

Echlin, P., & Hyde, P. J. W. (1972). The rationale and mode of application of thin films to non-conducting materials. *Scanning Electron Microscopy, 1*, 137–146.

Echlin, P., & Kayes, G. (1979). Thin films for high resolution conventional scanning electron microscopy. *Scanning Electron Microscopy, 2*, 21–30.

Edanaga, M. (1974). A scanning electron microscopic study of the endothelium of vessels. I. Fine structure of the endothelial surface of aorta and some other arteries in normal rabbits. *Archives Histologicum Japonicum, 3*, 1–14.

Eins, S., & Wilhems, E. (1976). Assessment of preparative volume changes in central nervous tissue using automatic image analysis. *Microscopie, 24*, 29–38.

Ekerot, C. F. (1985). Climbing fiber actions of Purkinje cells/plateau potential and long/lasting depression of parallel fibre responses. In J. R. Bloedel, J. Dichgans, & W. Precht (Eds.), *Cerebellar functions* (pp. 268–274). New York: Springer-Verlag.

Estable, C. (1923). Notes sur la structure comparative de l'écorce cérébelleuse, et dérivées physiologiques possibles. *Trabajos de los Laboratorios de Investigaciones Biológicas. (Madrid), 21*, 169–265.

Estable-Puig, R. F., & Estable-Puig, J. F. (1975). Brain cyst formation: A technique for SEM study of the central nervous system. *Scanning Electron Microscopy, 1*, 282–286.

Everhart, T. E., & Chung, M. S. (1972). Idealized spatial emission distribution of secondary electrons. *Journal of Applied Physic, 43*, 3307–3311.

Falcone, L. (1893). *La corteccia del cerveletto*. Napoli.

Fañanás, J., & Ramón y Cajal, S. (1916). Contribución al estudio de la neuroglia del cerebelo. *Trabajos de los Laboratorios de Investigaciones Biológicas. (Madrid), 14*, 163–179.

Fasekas, I., Bacsy, E., & Rappay, G. (1978). Identification of epithelial cells and fibroblast in hypophysis intermediate lobe cultures by scanning electron microscopy. *Acta Biológica Scientifica Hungárica, 29*, 407–416.

Feirabend, H. K., Choufoer, H., & Voogd, J. (1996). White matter of the cerebellum of the chicken (Gallus domesticus) : A quantitative light and electron microscopic analysis of myelinated fibers and fiber compartments. *Journal of Comparative Neurology, 369*, 236–251.

Fox, C. A. (1959). The intermediate cells of Lugaro in the cerebellar cortex of the monkey. *Journal of Comparative Neurology, 112*, 39–51.

Fox, C. A. (1962). Fine structure of the cerebellar cortex. In E. C. Crosby, T. Humphreys, & E. W. Lauer (Eds.), *Correlative anatomy of the nervous system* (pp. 192–198). New York: Mac Millan.

Fox, C. A., & Barnard, J. W. (1957). A quantitative study of the Purkinje cell dendritic branchlets and their relationship to afferent fibers. *Journal of Anatomy. (London), 91*, 299–313.

Fox, C. A., & Bertram, E. G. (1954). Connections of the Golgi cells and the intermediate cells of Lugaro in the cerebellar cortex of the monkey. *Anatomical Record, 118*, 423–424.

Fox, C. A., Siegesmund, K. A., & Dutta, C. R. (1964). The Purkinje cell dendritic branchlets and their relation with the parallel fibers: Light and electron microscopic observations. In M. M. Cohen, & R. S. Snider (Eds.), *Morphological and biochemical correlates of neural activity* (pp. 1112–1141). New York: Hoeber-Harper/Row.

Fox, C. A., Hillman, D. E., Siegesmund, K. A., & Dutta, C. R. (1967a). The primate cerebellar cortex: A Golgi and electron microscopic study. *Progress in Brain Research, 25*, 174–225.

Fox, C. A., Andrade, A., & Schwyn, R. C. (1967b). Climbing fiber branching in the granular layer. In R. Llinás (Ed.), *Neurobiology of cerebellar evolution and development* (pp. 603–611). Chicago: AMA/ERF. Institute For Biomedical Research.

Fujimoto, S., Yamamato, K., & Takeshige, Y. (1975). Electron microscopy of endothelial microvilli of large arteries. *Anatomical Record, 183*, 259–266.

Fusari, R. (1883). Sull' origine delle fibre nervose nello strato molecolare delle circunvoluzione cerebellari dell' uomo. *Attiguo Reale Accademia Scientifica. (Torino), 19*, 47–51.

Fusari, R. (1887). Untersuchungen über die feinere Anatomy des Gehirns der Teleostier. *International Monasschrift für Histologie und Physiologie, Bd. IV.*

Galbavy, E. S. S., & Olson, M. D. (1979). Morphogenesis of rod cells in the retina of the albino rat: A scanning electron microscopic study. *Anatomical Record, 195*, 707–718.

Gamliel, H. (1985). Optimum fixation conditions may allow air drying of soft biological specimens with minimum cell shrinkage and maximum preservation of surface features. *Scanning Electron Microscopy, 4*, 1649–1664.

Garcia-Segura, L. M., & Perrelet, A. (1984). Postsynaptic membrane domains in the molecular layer of the cerebellum: A correlation between presynaptic inputs and postsynaptic plasma membrane organization. *Brain Research, 32*, 255–256.

Garwicz, M. (1997). Sagittal organization of climbing fibre input to the cerebellar anterior lobe of the ferret. *Experimental Brain Research, 117*, 389–398.

Gertz, S. D., Rennels, M. L., Forbes, M. S., & Nelson, E. (1975). Preparation of vascular endothelium for scanning electron microscopy: A comparison of the effects of perfusion and immersion fixation. *Journal of Microscopy, 105*, 309–313.

Gierke, H. (1886). Die Stützsubstanz des Centralnervensystems. II. *Theil Archiv für mikroskopische Anatomie, 26*, 129–228.

Gobel, S. (1971). Axo-axonic septate junctions in the basket formations of cat cerebellar cortex. *Journal of Cell Biology, 51*, 328–333.

Golgi, C. (1874). Sulla fina anatomia del cerveletto umano. In *Opera Omnia. Vol. 1. Istologia Normale* (pp. 99–111). Istituto Lombardo di Sci. et Lett 8. Jan. Ch. V. Milan: Ulrico Hoepli 1903.

Golgi, C. (1882). Sulla fina anatomia degli organi centrali del sistema nervoso. I. Nota preliminari sulla struttura, morfología e vicendevoli rapporti delle cellule ganglionare. *Rivista Sperimentale de Freniatria, 8*, 165–195.

Golgi, C. (1883). Sulla fina anatomia degli organi centrali del sistema nervoso. IV. Sulla fina anatomia delle circunvoluzione cerebellari. *Rivista Sperimentale di Freniatria, 9*, 1–17.

Golgi, C. (1885). Sulla fina anatomia degli organi centrali del sistema nervoso. VIII. Tessuto interstiziale degli organi nervosi centrali (neuroglia). *Rivista Sperimentale di Freniatria, 11*, 72–123.

Gray, E. G. (1961). The granule cells, mossy synapses and Purkinje spine synapses of the cerebellum: Light and electron microscope observations. *Journal of Anatomy, 95*, 345–356.

Groenwegen, H. J., & Voogd, J. (1977). The parasagittal zonation within the olivo-cerebellar projections. I. Climbing fiber distribution in the vermis of cat cerebellum. Journal of *Comparative Neurology, 174*, 417–488.

Groniowsky, J., Wiczyskowa, W., & Walski, M. (1971). Scanning electron microscopic observations on the surface of vascular endothelium. *Folia Histochemica et Cytochemica, 9*, 243–246.

Gross, D. K., & De Boni, U. (1990). Colloidal gold labeling of intracellular ligands in dorsal root sensory neurons, visualized by scanning electron microscopy. *Journal of Histochemistry and Cytochemistry, 38*, 775–784.

Grovas, A. C., & O'Shea, K. S. (1984). A SEM examination of granule cell migration in the mouse cerebellum. *Journal of Neuroscience Research, 12*, 1–14.

Gundappa-Sulur, G., De Schutter, E., & Bower, J. M. (1999). Ascending granule cell axon: An important component of cerebellar cortical circuitry. *Journal of Comparative Neurology, 408*, 580–596.

Haggis, G. H. (1970). Cryofracture of biological material. *Scanning Electron Microscopy, 1*, 97–104.

Haggis, G. H., & Phipps-Todd, B. (1977). Freeze-fracture scanning electron microscopy. *Journal of Microscopy, 111*, 193–201.

Haggis, G. H., Bond, E. F., & Phipps-Todd, B. (1976). Visualization of mitochondrial cristae and nuclear chromatin by SEM. *Scanning Electron Microscopy, 1*, 282–286.

Hámori, J. (1964). Identification in the cerebellar isles of Golgi II axons endings by aid of experimental degeneration. In T. Tilbach (Ed.), *Electron microscopy* (pp. 291–292). Proceedings of theThird European Regional Conference on Electron Microscopy. Prague. Czechoslovak Academy of Science.

Hámori, J., & Szentágothai, J. (1964). The "crossing-over" synapses: An electron microscope study of the molecular layer on the cerebellar cortex. *Acta Biologica Academia Scientifica. Hungarica, 15*, 95–117.

Hámori, J., & Szentagothai, J. (1965). The Purkinje cell baskets: Ultrastructure of an inhibitory sinapse. *Acta Biologica Hungarica, 15*, 465–479.

Hámori, J., & Szentágothai, J. (1966a). Identification under the electron microscope of climbing fibers and their synaptic contacts. *Experimental Brain Research, 1*, 65–81.

Hámori, J., & Szentagothai, J. (1966b). Participation of Golgi neuron processes in the cerebellar glomeruli : An electron microscopic study. *Experimental Brain Research, 2*, 35–48.

Hámori, J., & Szentágothai, J. (1968). Identification of synapses formed in the cerebellar cortex by Purkinje axon collaterals: An electron microscopic study. *Experimental Brain Research, 15*, 118–128.

Hámori, J., Jakab, R. L., & Takacs, J. (1997). Morphogenetic plasticity of neuronal elements in cerebellar glomeruli during differentiation-induced synaptic reorganization. *Journal of Neural Transplantation and Plasticity, 6*, 11–20.

Hanke, S., & Reichenbach, A. (1987). Quantitative-morphometric aspects of Bergmann glial (Golgi epithelial) cell development in rats. A Golgi study. *Anatomy and Embryology, 177*, 183–188.

Hansen, C., & Linden, D. J. (2000). Long-term depression of the cerebellar climbing fiber-Purkinje neuron synapse. *Neuron, 26*, 473–482.

Hansson, H. A. (1970). Scanning electron microscopy of the rat retina. *Zeitschrift für Zellforschung, 107*, 23–44.

Hartwig, H. G., & Korf, H. W. (1978). The epiphysis cerebri of poikilothermic vertebrates: a photosensitive neuro-endocrine circumventricular organ. *Scanning Electron Microscopy, 2*, 163–168.

Harvey, R. J., & Napper, R. M. A. (1988). Quantitative study of granule and Purkinje cells in the cerebellar cortex of the rat. *Journal of Comparative Neurology, 274*, 151–157.

Hawkes, R., &, Gravel, C. (1991). The modular cerebellum. *Progress in Neurobiology, 36*, 309–327.

Hawkes, R., Brochin, G., Dore, L., Gravel, C., & Leclere, W. (1992). Zebrins: Molecular markers of compartmentation. In R. Llinás, & C. Sotelo (Eds.), *The cerebellum revisited.* New York: Springer-Verlag.

Heavner, J. E., Coates, P. W., & Pacz, G. (2000). Myelinated fibers of spinal cord blood vessels-sensory innervation?. *Current Review in Pain, 4*, 353–355.

Heinsen, H., & Heinsen, Y. L. (1983). Cerebellar capillaries. Qualitative and quantitative observations in young and senile rats. *Anatomy and Embryology, 168*, 101–116.

Held, H. (1897). Beiträge zur Struhtur der Nervenzellen und ihrer Fortzätze III. *Archiv für Anatomie und Physiologie. Anatomische Abtleitung. (Supplement)*, 273–312.

Herndon, R. M. (1963). The fine structure of Purkinje cell. *Journal of Cell Biology, 18*, 167–180.

Herndon, R. M. (1964). The fine structure of rat cerebellum. II.The stellate neurons, granule cells and glia. *Journal of Cell Biology, 23*, 277–293.

Hesslow, G., Svensson, P., & Ivarsson, M. (1999). Learned movements elicited by direct stimulation of cerebellar mossy fiber. *Neuron, 24*, 179–185.

Hillman, D. E. (1969). Neuronal organization of the cerebellar cortex in amphibia and reptilia. In R. Llinás (Ed.), *Neurobiology of cerebellar evolution and development* (pp. 279–325). Chicago: AMA/ERF. Institute For Biomedical Research.

Hillman, D. E., & Chen, S. (1981). Plasticity of synaptic size with constancy of total synaptic contact area on Purkinje cells in the cerebellum. *Progress in Clinical and Biological Research, 59 A*, 229–245.

Hirano, A., Cervos-Navarro, J., & Ohsugi, T. (1976). Capillaries in the subarachnoid space. *Acta Neuropathologica, 34*, 81–85.

Hirokawa, N. (1989). The arrangement of actin filaments in the postsynaptic cytoplasm of the cerebellar cortex revealed by quick-freeeze deep-etch electron microscopy. *Neuroscience Research, 6*, 269–275.

Hojo, J. (1994). An experimental scanning electron microscopic study of human cerebellar cortex using the t-butyl alcohol freeze-drying device. *Scanning Microscopy, 8*, 303–313.

Hojo, J. (1996). Specimen preparation of the human cerebellar cortex for scanning electron microscopy using a t-butyl alcohol freeze-drying device. *Scanning Microscopy (Supplement), 10*, 345–348.

Hojo, T. (1998). Human cerebellar cortex studied in scanning electron microscopy. *Scanning, 20*, 200–201.

Humphreys, W. J., Spurlock, B. O., & Johnston, J. S. (1974). Critical point drying of ethanol-infiltrated cryofracture biological specimens for scanning electron microscopy. *Scanning Electron Microscopy, 1*, 276–282.

Humphreys, W. J., Spurlock, B. O., & Johnston, J. S. (1975). Transmission electron microscopy of tissue prepared for scanning electron microscopy by ethanol-cryofracturing. *Stain Technology, 50*, 119–125.

Hunt, C. J., Taylor, M. J., & Pegg, D. E. (1982). Freeze-substitution and isothermal freeze-fixation studies to elucidate the pattern of ice formation in smooth muscle at 252 K (−21°C). *Journal of Microscopy, 125*, 177–186.

Ito, M. (1984). The *cerebellum and neural control* (pp. 21–115). New York: Raven Press.

Jaarsma, D., Diño, M. R., Cozzari, C., & Mugnaini, E. (1996). Cerebellar choline acetyl-transferase positive mossy fibres and their granule and unipolar brush cell targets: A model for central cholinergic nicotinic neurotransmission. *Journal of Neurocytology, 225*, 829–842.

Jakab, R. L., & Hámori, J. (1988). Quantitative morphology and synaptology of cerebellar glomeruli in the rat. *Anatomy and Embryology, 179*, 81–88.

Jakab, R. L. (1989). Three-dimensional reconstruction and synaptic architecture of cerebellar glomeruli in the rat. *Acta Morphologica Hungarica, 37*, 11–20.

Jakob, A. (1928). Das Kleinhirn. In W. V. Mollendorff (Ed.), *Handbuch der Mikroskopischen Anatomie des Menschen. Vol. IV* (pp. 771–831). Berlin: Julius Springer.

Jansen, J., & Brodal, A. (1958). Das Kleinhirn. In W. Wargman (Ed.), *Handbuch der Mikroskopischen Anatomie des Menschen. Bd IV/8* (pp. 101–103). Berlin: Springer.

Ji, J., & Hawkes, R. (1994). Topography of Purkinje cell compartments and mossy fiber terminal fields in lobules II and III of the rat cerebellar cortex: Spinocerebellar and cuneocerebellar projections. *Neuroscience, 61*, 935–954.

Kanaseki, T., Ykeuchi, Y., & Tashiro, J. (1998). Rough surfaced smooth endoplasmic reticulum in rat and mouse Purkinje cells visualized by quick-freezing techniques. *Cell Structure and Function, 23*, 373–387.

Katora, M. E., & Hollis, T. M. (1976). Regional variation in rat aortic endothelial surface morphology: Relationship to regional aortic permeability. *Journal of Experimental Molecular Pathology, 24*, 23–34.

Kawamura, J., Gertz, S. D., Sunaga, T., Rennels, M. L., & Nelson, E. (1974). Scanning electron microscopic observations of the luminal surface of the rabbit common carotid artery subjected to ischemia by arterial occlusion. *Stroke, 5*, 765–774.

Kenyon, G. T. (1997). A model of long-term memory storage in the cerebellar cortex: A possible role for plasticity at parallel fiber synapses onto stellate/basket interneurons. *Proceedings of the National Academy of Science of USA, 94*, 14200–14205.

Kerr, W. H., & Bishop, G. A. (1991). Topographical organization in the origin of serotoninergic projections to different regions of the cat cerebellar cortex. *Journal of Comparative Neurology, 304*, 502–515.

Kessel, R. C., & Kardon, R. H. (1981). Scanning electron microscopy of mammalian neuroepithelia. *Biomedical Research (Supplement), 2*, 483–489.

Khan, M. A. (1993). Histochemical and ultrastructural investigation of heterogeneous Purkinje neurons in mammalian cerebellum. *Cellular and Molecular Biology Research, 39*, 789–795.

Kido, T., & Sekitani, T. (1994). High resolution scanning electron microscopic study of cell organelles of the chick's vestibular ganglion. *European Archives of Oto-Rhino-Laryngology, 51*, 1–47.

Kido, T., Sekitani, T., Okami, K., Endo, S., & Moriya, K. (1993). Ultrastructure of the chick vestibular ganglion and vestibular nucleus. A scanning electron microscopic study. *Acta Oto-Laryngologica (Supplement), 503*, 161–165.

Kim, S. U., Kim, K. M., Moretto, G., & Kim, J. H. (1985). The growth of fetal human sensory ganglion neurons in culture: A scanning electron microscopic study. *Scanning Electron Microscopy, 2*, 843–848.

King, J. S., Chen, Y. F., & Bishop, G. A. (1993). An analysis of HRP-filled basket cell axons in the cat's cerebellum. II. Axonal distribution. *Anatomy and Embryology (Berlin), 188*, 299–305.

Kleim, J. A., Swain, R. A., Armstrong, K. A., Napper, R. M., Jones, T. A., & Greenough, W. T. (1998). Selective synaptic plasticity within the cerebellar cortex following complex motor skill learning. *Neurobiology of Learning and Memory, 69*, 274–289.

Kleiter, N., & Lametschwandtner, A. (1995). Microvascularization of the cerebellum in the turtle, *Pseudemys Scripta Elegans* (Reptilia). A scanning electron microscope study of microvascular corrosion casts, including stereological measurements. *Anatomy and Embryology, 191*, 145–153.

Kölliker, A. (1890). Zur feineren Anatomie des Zentralen Nervensystems. I. Das Kleinhirn. *Zeitschrift für Wissenschartliche Zoologie, 49*, 663–689.

Kosaka, T., Kosaka, K., Nakayama, T., Hunziker, W., & Heiszmann, C. W. (1993). Axons and axon terminals of cerebellar Purkinje cells and basket cells have higher levels of parvalbumin immunoreactivity than somata and dendrites: Quantitative analysis by immunogold labeling. *Experimental Brain Research, 93*, 483–491.

Kotskovich, R. P. (1981). Interaction of Purkinje cells, capillaries and glia in the cerebellar cortex of normal and hypokinetic cats. *Neuchnye doklady visshe shkoly Biologicheskic Nauki, 1*, 50–54.

Kril, J. J., Flowers, D., & Butterworth, R. F. (1997). Distinctive pattern of Bergmann glial pathology in human hepatic encephalopathy. *Molecular and Chemical Neuropathology, 3*, 279–287.

Krstic, R. V. (1974). Scanning electron microscopic study of the freeze-fractured pineal body of the rat. *Cell and Tissue Research, 209*, 129–135.

Kugler, P., & Drenckhahn, D. (1996). Astrocytes and Bergmann glia as an important site of nitric oxide synthase I. *Glia, 16*, 165–173.

Lache, C. (1906). Sur les corbeilles des cellules de Purkinje. *Compte Rendu de la Société. Biologique (Paris), 9*, 30–39.

Laine, J., & Axelrad, H. (1996). Morphology of the Golgi-impregnated Lugaro cell in the rat cerebellar cortex: A reappraisal with a description of its axon. *Journal of Comparative Neurology*, 375, 618–640.

Laine, J., & Axelrad, H. (1998). Lugaro cells target basket and stellate cells in the cerebellar cortex. *Neuroreport, 9*, 2399–2403.

Lafarga, M., Andres, M. A., Calle, E., & Berciano, M. T. (1998). Reactive gliosis of immature Bergmann glia and microglial cell activation in response to cell death of granule cell precursors induced by methylazoxymethanol treatment in developing rat cerebellum. *Anatomy and Embryology, 198*, 111–112.

Lametschwandtner, A., Simonsberg, O., & Adam, H. (1977). Vascularization of the pars intermedia of the hypophysis in the toad, Bufo Bufo (L.) (Amphibia, Anura). *Cell and Tissue Research, 179*, 11–16.

Landis, D., & Reese, R. S. (1974). Differences in membrane structure between excitatory and inhibitory synapses in the cerebellar cortex. *Journal of Comparative Neurology, 155*, 93–126.

Lange, W. (1977). Comparative studies on cell density and vascular density in the cerebellar cortex. *Verhanlungen der Anatomische Gesellschaft, 71*, 1021–1022.

Lange, W., & Halata, Z. (1972). Ultrastructure of capillaries in the cerebellar cortex and the pericapillary compartment. *Zeitschrift für Zellforschung und Mikroskopische Anatomie, 128*, 83–99.

Lange, W., & Halata, Z. (1979). Comparative studies on the pre- and post-terminal blood vessels in the cerebellar cortex of Rhesus monkey, cat and rat. *Anatomy and Embryology, 158*, 51–62.

Larina, V. N. (1980). Certain features of the vascular-capillary network of the cerebellar cortex of dogs. *Arkchiv Anatomi, Gistologii i Embriologii, 78*, 51–54.

Larsell, O., & Jansen, J. (1972). *The comparative anatomy and histology of the cerebellum. The human cerebellum, cerebellar connections and cerebellar cortex* (pp. 3–197). Minneapolis: University of Minnesota Press.

Larramendi, L. M. H., & Victor, T. (1967). Synapses on the Purkinje cell spines in the mouse. An electron microscopic study. *Brain Research, 5*, 15–30.

Lazzari, M., & Franceschini, V. (2000). Structural and spatial organisation of brain parenchymal vessels in the lizard, *Podarcis Sicula*: A light, transmission and scanning electron microscopy study. *Journal of Anatomy, 197*, 167–175.

Lee, R. M. K. W. (1983). A critical appraisal of the effects of fixation, dehydration and embedding on cell volume. In J. P. Revel, T. Bernard, & G. H. Haggis (Eds.), *The science of biological specimen preparation for microscopy and microanalysis* (pp. 61–70). Chicago: Scanning Microscopy International.

Lee, M. M. L., & Chieng, S. (1979). Morphological effects of pressure changes on canine carotid artery endothelium as observed by scanning electron microscopy. *Anatomical Record, 194*, 1–4.

Lee, R. M. K. W., Mc Kenzie, R., Kobayashi, K., Garfield, R. E., Forrest, J. B., & Daniel, E. E. (1982). Effects of glutaraldehyde fixative osmolarities on smooth muscle cell volume and osmotic reactivity of the cells after fixation. *Journal of Microscopy, 125*, 77–88.

Lemkey-Johnston, N., & Larramendi, L. M. H. (1968a). Morphological characteristics of mouse stellate and basket cells and their neuroglial envelope: An electron microscopic study. *Journal of Comparative Neurology, 134*, 39–72.

Lemkey-Johnston, N., & Larramendi, L. M. H. (1968b). Types and distribution of synapses upon basket and stellate cells of the mouse cerebellum: An electron microscopic study. *Journal of Comparative Neurology, 134*, 73–112.

Lewis, R. E. (1971). Studying neuronal architecture and organization with the scanning electron microscope. In O. Johari, & I. Corvin (Eds.), *Scanning electron microscopy,*(pp. 283–288). Chicago: IITRI.

Lewis, R. E., & Nemanic, K. M. K. (1973). Critical point drying techniques. *Scanning Electron Microscopy, 2*, 767–774.

Livesey, F. J., & Fraher, J. P. (1992). Experimental traction injuries of cervical spinal nerve roots: A scanning EM study of rupture patterns in fresh tissue. *Neuropathology and Applied Neurobiology, 18*, 376–386.

Llinás, R. (1984). Functional significance of the basic cerebellar circuit in matter coordination. In J. R. Bloedel, J. Dichgans, & W. Precht (Eds.), *Cerebellar functions* (pp. 170–185). New York: Springer-Verlag.

Llinás, R., & Sasaki, K. (1989). The functional organization of the olivocerebellar system as examined by a multiple Purkinje cell recording. *European Journal of Neuroscience, 1*, 587–602.

Löfberg, J., Ahlfors, K., & Fallstrom, C. (1980). Neural crest cell migration in relation to extracellular matrix organization in the embryonic axolotl trunk. *Developmental Biology, 75*, 148–167.

Low, F. N. (1982). The central nervous system in scanning electron microscopy. *Scanning Electron Microscopy, 2*, 869–890.

Low, F. N. (1989). Microdissection by ultrasonication for scanning electron microscopy. In P. M. Motta (Ed.), *Cells and tissue: A three-dimensional approach by modern techniques in microscopy* (pp. 571–580). New York: Alan R. Liss.

Luft, J. H. (1971). Ruthenium red and violet. II. Fine structural localization in animal tissue. *Anatomical Record, 171*, 369–416.

Lugaro, E. (1894). Sulle connessioni tra gli elementi nervosi della corteccia cerebellare con considerazioni generali sul significato fisiologico dei repporti tra gli elementi nervosi. *Rivista. Sperimentale di Freniatria, 20*, 297–331.

Luget, B. J. (1970). Physical changes occurring in frozen solutions during rewarming and melting. In G. E. W. Wolstenholme, & M. O'Connor (Eds.), *The frozen cell* (pp. 27–50), London: J. & J. Churchill.

Mann-Metzer, P., & Yarom, Y. (2000). Electronic coupling synchronizes interneuron activity in the cerebellar cortex. In N. M. Gerrits, T. J .H. Ruigrok, & C. I. De Zeeuw (Eds.), *Cerebellar modules: Molecules, morphology and function* (pp. 115–122). Progress in Brain. Research. Vol. 124. New York: Elsevier.

Martone, M. E., Zhang, I., Simpliciano, M., Carrager, B. O., & Ellisman, M. (1993). Three-dimensional visualization of the smooth endoplasmic reticulum in Purkinje cell dendrites. *Journal of Neuroscience, 13*, 4636–4646.

Massion, J. (1993). Major anatomical functional relations in the cerebellum. *Revue Neurologique (Paris), 149*, 600–606.

Mathew, T. C. (1998). Supraependymal neuronal elements of the floor of the fourth ventricle in adult rat: A scanning electron microscopic study. *Journal of Submicroscopic Cytology and Pathology, 3*, 175–181.

Matonoha, P., & Zechmaister, A. (1978). Scanning electron microscopic observations on intimal surface of normal and atherosclerotic arteries. *Acta Morphologica Accademia Scientifica Hungarica, 26*, 173–184.

Matsuda, S., & Uehara, Y. (1981). Cytoarchitecture of the rat dorsal root ganglia as revealed by scanning electron microscopy. *Journal of Electron Microscopy, 30*, 136–140.

Matsumura, A., & Kohno, K. (1991). Microtubule bundles in fish cerebellar Purkinje cells. *Anatomy and Embriology (Berlin), 183*, 105–110.

McAuliffe, W. G., & Hess, A. (1990). Human chromogranin A-like immunoreactivity in the Bergmann glia of the rat brain. *Glia, 3*, 13–16.

Meek, J., & Nieuwenhuys, R. (1991). Palisade pattern of mormyrid Purkinje cells: A correlated light and electron microscopic study. *Journal of Comparative Neurology, 306*, 156–192.

Melik-Musyan, A. B., & Fanardzhyan, V. V. (1998). Structural organization and Lugaro neuron connections in the cat cerebellar cortex. *Morphologiia, 113*, 44–48

Meller, K. (1987). The cytoskeleton of cryofixed Purkinje cells of the chicken cerebellum. *Cell and Tissue Research, 247*, 155–165.

Meller, S., & Denis, B. J. (1993). A scanning and transmission electron microscopic analysis of the cerebral aqueduct in the rabbit. *Anatomical Record, 237*, 124–140.

Merchant, R. E. (1979). Scanning electron microscopy of spinal nerve exits of normal and BCG infected dogs. *Scanning Electron Microscopy, 3*, 341–345.

Messer, A. (1977). The maintenance and identification of mouse cerebellar granule cells in monolayer culture. *Brain Research, 130*, 1–12.

Mestres, P., & Rascher, K. (1994). The ventricular system of the pigeon brain: A scanning electron microscope study. *Journal of Anatomy, 184*, 35–58.

Monteiro, R. A. (1989). Morphometric differences between basket cells and stellate cells of rat neocerebellum (Crus I and Crus II). *Journal of Submicroscopic Cytology and Pathology, 21*, 725–736.

Monteiro, R. A., Rocha, E., & Marini-Abreu, M. M. (1994). Heterogeneity and death of Purkinje cells of rat neocerebellum (Crus I and Crus II): Hypothetic mechanisms based on qualitative and quantitative microscopical data. *Journal für Hirnforschung, 35*, 205–222.

Mugnaini, E. (1972). The histology and cytology of the cerebellar cortex. In O. Larsell, & J. Jansen (Eds.), *The comparative anatomy and histology of the cerebellum. The human cerebellum, cerebellar connections and cerebellar cortex* (pp. 201–251). Minneapolis: University of Minnesota Press.

Mugnaini, E., & Floris, A. (1994). The unipolar brush cell: A neglected neuron of the mammalian cortex. *Journal of Comparative Neurology, 339*, 174–180.

Mugnaini, E., Atluri, R. L., & Hank, J. C. (1974). Fine structure of granular layer in turtle cerebellum with special emphasis on large glomeruli. *Journal of Neurophysiology, 37*, 1–29.

Mugnaini, E., Floris, A., & Wright-Gross, M. (1994). The extraordinary synapses of unipolar brush cells: An electron microscopic study in the rat cerebellum. *Synapse, 16*, 284–311.

Mugnaini, E., Diño, M. R., & Jaarsma, D. (1997). The unipolar brush cells of the mammalian cerebellum and cochlear nucleus: Cytology and microcircuitry. In C. I. De Zeeuw, P. Strata, & J. Voogd (Eds.), *The cerebellum: From structure to control* (pp. 131–150). Progress in Brain Research. Vol. 114. Amsterdam: Elsevier.

Muller, T., Moller, T., Neuhaus, J., & Kettenmann, H. (1996). Electrical coupling among Bergmann glial cells and its modulation by glutamate receptor activation. *Glia, 17*, 274–284.

Muñoz, D. G. (1990). Monodendritic neurons: A cell type in the human cerebellar cortex identified by chromogranin A-like immunoreactivity. *Brain Research, 528*, 335–338.

Murphy, M. G., O'Leary, J. L., & Corntlath, D. (1973). Axoplasmic flow in cerebellar mossy and climbing fibers. *Archives of Neurology, 28*, 118–130.

Nemanic, M. K. (1972). Critical point drying, cryofracture and serial sectioning. *Scanning Electron Microscopy, 1*, 297–304.

Negyessi, L., Vidnyanszky, Z., Kuhn, R., Knopfel, T., Gores, T. J., & Hamori, J. (1997). Light and electron microscopic demonstration of mGlur5 metabotropic glutamate receptor immunoreactive neuronal elements in the rat cerebellar cortex. *Journal of Comparative Neurology, 385*, 641–650.

Nousek-Goebel, N. A., & Press, M. F. (1986). Golgi-electron microscopic study of sprouting endothelial cells in the neonatal rat cerebellar cortex. *Brain Research, 395*, 67–73.

O'Donohue, D. L., King, J. S., & Bishop, G. A. (1989). Physiological and anatomical studies of the interactions between Purkinje cells and basket cells in the cat's cerebellar cortex: Evidence for a unitary relationship. *Journal of Neuroscience, 9*, 2141–2150.

O'Leary, J. L., Petty, J. M., Smith, M. B., & Inukai, J. (1968). Cerebellar cortex of rat and other animals. An ultrastructural study. *Journal of Comparative Neurology, 134*, 401–432.

O'Leary, J. L., Inukai, F., & Smith, M. B. (1971). Histogenesis of the cerebellar climbing fiber in the rat. *Journal of Comparative Neurology, 142*, 377–392.

Obersteiner, H. (1888). Anleitung beim Studium des Baues der Nervösen Zentralorgane. *Arbeits der Neurologische Institut.* Wien University, *22*, 325–335.

Obsuki, K. (1972). Scanning electron microscopic studies on rabbit's spinal cord by resin cracking method. *Archives Histologicum Japonicum, 34*, 405–415.

Okhotin, V. E., & Kalinichenko, S. G. (1999). Localization of NO-synthase in Lugaro cells and mechanisms of NOergic interactions between inhibitory interneurons of rabbit cerebellar cortex. *Morfologiia, 115*, 52–61.

Oscarsson, O. (1969). The sagittal organization of the cerebellar anterior lobe as revealed by the projections patterns of the climbing fiber system. In R. Llinas (Ed.), *Neurobiology of cerebellar evolution and development* (pp. 525–537). Chicago: AMA-ERF.

Oscarsson, O. (1979). Functional units of the cerebellum sagittal zones and microzones. *Trends in Neuroscience, 2*, 143–145.

Paine, C. F., & Low, F. N. (1975). Scanning electron microscopy of cardiac endothelium of the dog. *American Journal of Anatomy, 142*, 137–158.

Palay, S. L. (1956). Synapses in the central nervous system. *Journal of Biophysics and Biochemical Cytology (Supplement), 2*, 193–202.

Palay, S. L., & Chan-Palay, V. (1974). *Cerebellar cortex. Cytology and organization* (pp. 1–348). Berlin: Springer-Verlag.

Palay, S. L., & Palade, G. E. (1955). The fine structure of neurons. *Journal of Biophysics and Biochemical Cytology, 1*, 69–88.

Palkovitz, M., Magyar, P., & Szentágothai, J. (1971). Quantitative histological analysis of the cerebellar cortex in the cat. III. Structural organization of the molecular layer. *Brain Research, 34*, 1–18.

Pappenheimer, J. R. (1953). Passage of molecule through capillary walls. *Physiological Rewiews, 33*, 387–423.

Pastukhov, V. A. (1974). Interrelations of capillaries and neurons in different formations of the brain and cerebellum. *Fiziologicheskii Zhurnal Sssr Iment M. Schenova, 60*, 1423–1427.

Paul, L. A., Fried, L., & Scheibel, A. B. (1984). Scanning electron microscopy of the central nervous system. II. The hippocampus. *Brain Reserach Reviews, 8*, 177–192.

Pensa, A. (1931). Osservazioni e considerazioni sulla structtura della corteccia cerebellare dei mammiferi. *Reale Accademia Nazionale dei Lincei. Ser. VI, 5*, 1–26.

Peters, A., Palay, S. L., & Webster, H. D. F. (1970). The *fine structure of the nervous system. The cells and their processes* (pp. 118–119). New York: Harper/Row.

Peters, K. R. (1980). Improved handling of structural fragile cell-biological specimens during electron microscopic preparation by the exchange method. *Journal of Microscopy, 118*, 429–441.

Peters, K. R. (1985a). Noise reduction in high magnification micrographs by soft focus printing and digital image processing. *Scanning, 7*, 205–215.

Peters, K. R. (1985b). Working at higher magnifications in scanning electron microscopy with secondary and backscattered electrons on metal coated biological specimens and imaging macro molecular cell membrane structures. *Scanning Electron Microscopy, 4*, 1519–1544.

Phillips, M. I., Balhorn, L., Leavitt, M., & Hoffman, W. (1974). Scanning electron microscope study of the rat subfornical organ. *Brain Research, 80*, 95–100.

Ponti, U. (1897). Sulla corteccia cerebellare della cavia. *Monitor di Zoologia Italiano, 8*, 1–188.

Popoff, R. (1896). Weitere Beiträge zur Frage über dir Histogenese der Kleinhirnrinde. *Biologische Zentralblat, 15*, 745–752.

Purkinje, J. E. (1837). Neueste Untersuchungen aus der Nerven und Hirn Anatomie. In K. Stenberg, & J.V. Von Krombholtz (Eds.), *Bericht über die Versammlung deutscher Naturforscher und Aertze* (pp. 177–180). Prague University. Prague.

Ramón y Cajal, S. (1888). Estructura de los centros nerviosos de las aves. *Revista Trimestral. Histológica (Madrid), 1*, 1–10.

Ramón y Cajal, S. (1890a). A propos de certains eléments bipolaires du cervelet avec quelques détails nouveaux sur l'évolution des fibres cérébelleuses. *International Monattschrift für Anatomie und Physiologie, 7*, 447–468.

Ramón y Cajal, S. (1890b). Sur les fibres nerveuses de la couche granuleuse du cervelet et sur l'evolution des eléments cerebelleuses. *International Monattschrift für Anatomie und Physiologie, 7*, 12–30.

Ramón y Cajal, P. (1896). Las células estrelladas de la capa molecular del cerebelo de los reptiles. *Revista Trimestral. Micrográfica (Madrid), 1*, 149–150.

Ramón y Cajal, S. (1896a). El azul de metileno en los centros nerviosos. *Revista Trimestral Micrográfica (Madrid), 1*, 151–204.

Ramón y Cajal, S. (1911). *Histologie du système nerveux de l'homme et des vertébrés. Vol. I.* (pp. 73–105). (L. Azoulay, Trans.) Paris: Maloine.

Ramón y Cajal, S. (1926). Sur les fibres mosseuses et quelques points douteux de la texture de l'écorce cérébelleuse. *Trabajos del Laboratorio de Investigaciones Biológicas (Madrid), 24*, 215–251.

Ramón y Cajal, S. (1954). *Neuron theory or reticular theory? Objective evidence of the anatomical unity of nerve cells* (pp. 1–142). (M. Ubeda-Purkiss & C. A. Fox, Trans.), Madrid: Consejo Superior de Investigaciones Científicas, Instituto Ramón y Cajal.

Ramón y Cajal, S. (1955). *Histologie du système nerveux de l'homme et des vertébrés. Vol. I* (pp. 1–79). Madrid: Consejo Superior de Investigaciones Científicas. Instituto Ramón y Cajal.

Reese, T. S., & Karnovsky, M. J. (1967). Fine structural localization of a blood–brain barrier to exogenous peroxidase *Journal of Cell Biology, 34*, 2707–2717.

Reese, B. F., Landis, D. M., & Reese, T. S. (1985). Organization of the cerebellar cortex viewed by scanning electron microscopy. *Neuroscience, 14*, 133–146.

Reichenbach, A., Siegel, A., Rickmann, M., Wolff, J. R., Noone, D., & Robinson, S. R. (1995). Distribution of Bergmann glial somata and processes: Implications for function. *Journal für Hirnforschung, 36*, 509–517.

Reina, M. A., Dittmann, M., Lopez Garcia, A., & Van Zundert, A. (1997). New perspectives in the microscopic structure of human dura mater in the dorsolumbar region. *Regional Anesthesia, 22*, 161–166.

Retzius, G. (1892a). Die nervösen Elemente der Kleinhirnrinde. *Biologische Untersuchungen, N.F.Bd. III* (pp. 17–29). Stockholm: Samsom/Wallive.

Retzius, G. (1892b). Kleinere Mittheilungen von dem Gebiete der Nervenhistologie.I. Über die Golgi'schen Zellen und die Kletternfasern Ramón y Cajal's in der Kleinhinrinde. *Biologische. Untersuchungen. N. F. Bd. IV* (pp. 57–59). Stockholm: Samson/Wallive.

Retzius, G. (1892c). Die Neuroglia des Gehirns beim Menschen und beim Säugethieren. *Biologische Untersuchungen. N.F. Bd. III* (pp. 30–35). Stockholm: Samson/Wallive.

Retzius, G. (1894). Die Neuroglia des Gehirn beim Menschen und beim Säugethieren: Die Neuroglia des Kleinhirns. *Biologische Untersuchungen N.F. Bd. VI* (pp. 16–20). Jena: Gustav/Fischer.

Riesco, J. M., Juanes, J. A., Sanchez, F., Blanco, E., Carretero, J., Maillo, A., & Vazquez, R. (1993). Combined TEM and SEM analysis of the rostral wall of the human III ventricle. *Histology and Histopathology, 8*, 213–218.

Rivera-Domínguez, M., Mettler, F. A., & Novack, C. R. (1974). Origin of cerebellar climbing fibers in the Rhesus monkey. *Journal of Comparative Neurology, 155*, 331–342.

Rusakov, D. A., Podini, P., Villa, A., & Meldolesi, J. (1993). Tridimensional organization of Purkinje neuron cisternal stacks, as specialized endoplasmic reticulum subcompartments rich in inositol 1,4,5- trisphosphate receptors. *Journal of Neurocytology, 22*, 273–282.

Saetersdal, T. S., & Myklebust, R. (1975). Identification of nerve fibers in the scanning electron microscope. *Journal of Microscopy, 103*, 63–69.

Sahin, M., & Hockfield, S. (1990). Molecular identification of the Lugaro cell in the cat cerebellar cortex. *Journal of Comparative Neurology, 301*, 575–584.

Sanchez, A., Bilinski, M., González Nicolini, V., Villar, M. J., & Tramezzani, J. H. (1997). Galanin and cholecystokinin in cultured magnocellular neurons isolated from adult rat supraoptic nuclei: A correlative light and scanning electron microscopical study. *Histochemical Journal, 29*, 631–638.

Sarphie, T. G., Carey, M. E., Davidson, J. F., & Soblosky, J. S. (1999). Scanning electron microscopy of the floor of the fourth ventricle in rats subjected to graded impact injury to the sensorimotor cortex. *Journal Neurosurgery, 90*, 734–742.

Scheibel, M. E., & Scheibel, A. B. (1954). Observations on the intracortical relations of the climbing fibers of the cerebellum. *Journal of Comparative Neurology, 101*, 733–763.

Scheibel, A. B., Paul, L. A., & Fried, I. (1981). Scanning electron microscopy of the central nervous system. I. The cerebellum. *Brain Research Reviews, 3*, 207–228.

Schroeder, A. H. (1929). Die Gliaarchitecktonik des Menschlichen Kleinhirns. *Journal of Psychology and Neurology, 38*, 234–257.

Scott, D. E., Kozlowski, G. P., Paul W. K., Ramalingam, S., & Krobisch-Dudley, G. (1973). Scanning electron microscopy of the human cerebral ventricular system. II. The fourth ventricle. *Zeitschrift für Zellforschung, 139*, 61–68.

Seymour, R. M., & Berry, M. (1975). Scanning and transmission electron microscope studies of interkinetic nuclear migration in the cerebral vesicles of the rat. *Journal of Comparative Neurology, 160*, 105–125.

Schweighofer, N. (1998). A model of activity-dependent formation of cerebellar microzones. *Biological Cybernetics, 79*, 97–107.

Shelton, E., & Mowczko, W. E. (1978). Membrane blisters. A fixation artifact. A study of fixation for scanning electron microscopy. *Scanning, 1*, 166–173.

Shimamoto, T., Yamashita, Y., & Sunaga, T. (1969). Scanning electron microscopic observation of endothelial surface of hearth and blood vessel. *Proceedings of Japanese Academy of Science, 45*, 507–511.

Shinoda, Y., Sughihara, I., Wu, H. S., & Sugiuchi, Y. (2000). The entire trajectory of single climbing and mossy fibers in the cerebellar nuclei and cortex. In N. M. Gerrits, T. J. H. Ruigrok, & C.I. De Zeeuw (Eds.), *Cerebellar modules: Molecules, morphology and function*. Progress in Brain Research, *124*, 173–186.

Shotton, D. M., Heuser, J. E., Reese, B. F., & Reese, R. S. (1979). Postsynaptic membrane folds of the frog neuromuscular junction visualized by scanning electron microscopy. *Neuroscience, 4*, 427–435.

Siegel, A., Reichenbach, A., Hanke, S., Senitz, D., Brauer, K., & Smith, T. G., Jr. (1991). Comparative morphometry of Bergmann glial (Golgi epithelial) cells. A Golgi study. *Anatomy and Embryology, 183*, 605–612.

Siew, S. (1979). Scanning electron microscopy of the human spinal cord and dorsal root ganglia. *Scanning Electron Microscopy, 1979, 3*, 421–429.

Sievers, J., Mangold, U., & Berry, M. (1985). 6-OHDA-induced ectopia of external granule cells in the subarachnoid space covering the cerebellum. III. Morphology and synaptic organization of ectopic cerebellar neurons: A scanning and transmission electron microscopic study. *Journal of Comparative Neurology, 232*, 319–330.

Simionescu, N. (1980). Transcytosis and endocytosis in the endothelial cell. *Second International Congress of Cell Biology. European Journal of Cell Biology (Abstract), 22*, 180.

Simionescu, M., Simionescu, N., & Palade, G. E. (1974). Morphometric data on the endothelium of blood capillaries. *Journal of Cell Biology, 60*, 128–152.

Smith, V., Ryan, T. W., Michic, A. A., & Smith, D. S. (1971). Endothelial projections as revealed by scanning electron microscope. *Science, 173*, 925–927.

Smirnov, A. E. (1897). Über eine besondere von Nervenzellen der Molecularschicht des Kleinhirn bei erwachsenen Saügetieren und beim Menschen. *Anatomischer Anzeiger, 13*, 636–642.

Sotelo, C. (1969). Ultrastructural aspects of the cerebellar cortex of the frog. In R. Llinás (Ed.), *Neurobiology of cerebellar evolution and development* (pp. 327–37). Chicago: AMA/ERF.

Sotelo, C. (1970). Stellate cells and their synapses on Purkinje cells in the cerebellum of the frog. *Brain Research, 17*, 510–514.

Stewart, G. T., Ritchie, G. M., & Lynck, R. R. (1973). A scanning and transmission electron microscopic study of canine jugular veins. *Scanning Electron Microscopy, 1*, 473–480.

Still, W. J., & Dennison, S. (1974). The arterial endothelium of the hypertensive rat. A scanning and transmission electron microscopical study. *Archives of Pathology, 92*, 337–342.

Sturrock, R. R. (1990). A quantitative histological study of Golgi II neurons and pale cells in different cerebellar regions of the adult and ageing mouse brain. *Zheitschrift für Mikroskopische Anatomy Forschung, 104*, 705–714.

Suarez, I., Bodega, M., Arilla, E., Rubio, M., Villalba, R., & Fernández, B. (1992). Different response of astrocytes and Bergmann glial cells to portacaval shunt: An immunohistochemical study in the rat cerebellum. *Glia, 6*, 172–179.

Sultan, F. (2000). Exploring a critical parameter of timing in the mouse cerebellar microcircuitry: The parallel fiber diameter. *Neuroscience Letter, 280*, 41–44.

Sultan, F., & Bower, J. M. (1998). Quantitative Golgi study of the rat cerebellar molecular layer interneurons using principal component analysis. *Journal of Comparative Neurology, 393*, 353–373.

Sunaga, T., Shimamoto, T., & Nelson, E. (1973). Correlated scanning and transmission electron microscopy of arterial endothelium. *Scanning Electron Microscopy, 1*, 458–464.

Swinehart, P. A., Pently, D. L., & Kardong, K. W. (1976). Scanning electron microscopic study of the effects of pressure on the luminal surface of the rabbit aorta. *American Journal of Anatomy, 145*, 137–142.

Szentágothai, J., & Rajkovits, V. (1959). Über den Ursprung der Kletterfasern des Kleinhirns. *Zeitschrift Anatomische Entwicklungsgesch, 121*, 130–141.

Tachibana, S., Takeuchi, M., & Fujiwara, R. (1985). Visualization of autonomic varicose terminal axons by scanning electron microscopy. *Journal of Electron Microscopy, 34*, 136–138.

Takahashi-Iwanaga, H. (1992). Reticular endings of Purkinje cell axons in the rat cerebellar nuclei: Scanning electron microscopic observations of the pericellular plexus of Cajal. *Archives of Histology and Cytology, 55*, 307–314.

Takahashi-Iwanaga, H., Murakami, T., & Abe, K. (1998). Three-dimensional microanatomy of perineuronal proteoglycan nets enveloping motor neurons in the rat spinal cord. *Journal of Neurocytoly, 27*, 817–827.

Takumida, M., Miyawaki, H., Harada, Y., & Anniko, M. (1995). Three-dimensional organization of cytoskeletons in the vestibular sensory cells. *Journal of Otorhinolaryngoly and Related Specialities, 57*, 100–104.

Tamega, O. J., Tirapelli, L. F., & Petroni, S. (2000). Scanning electron microscopy study of the choroid plexus in the monkey (Cebus Apella Apella). *Archives of Neuropsiquiatry, 58*, 820–825.

Tanaka, K., Iino, A., & Naguro, T. (1976). Scanning electron microscopic observations on intracellular structures of ion-etched materials. *Archives Histologicum Japonicum, 39*, 165–175.

Tan, J., Gerritz, N. M., Nanhoe, R., Simpson, J. I., & Woogd, J. (1995). Zonal organization of the climbing fiber projections to the flocculus and nodulus of the rabbit: A combined axonal tracing and acetylcholinesterase histochemical study. *Journal of Comparative and Neurology, 356*, 23–50.

Tandler, C. J., Rios, H., & Pellegrino de Iraldi, A. (1997). Differential staining of two subpopulations of Purkinje neurons in rat cerebellum with acid dyes. *Biotechnic and Histochemistry, 72*, 231–239.

Terasaki, M., Slater, N. T., Feiw, A., Schmidek, A., & Reese, T. S. (1994). Continuous network of endoplasmic reticulum in cerebellar Purkinje neurons. *Proceedings of the National Academy of Sciences of USA, 91*, 7510–7514.

Terrazas, R. (1897). Notas sobre la neuroglia del cerebelo y el crecimiento de los elementos nerviosos. *Revista Trimestral Micrográfica (Madrid), 2*, 49–65.

Teune, T. M., Van Der Burg, J., De Zeeuw, C. I., Voogd, J., & Ruigrok, T. J. H. (1998). Single Purkinje cell can innervate multiple classes of projection neurons in the cerebellar nuclei of the rat: A light microscopic and ultrastructural triple tracer study. *Journal of Comparative Neurology, 392*, 164–168.

Teune, T. M., Van Der Burg, J., Van Der Moer, J., Woogd, J., & Ruigrok, T. J. H. (2000). Topography of cerebellar nuclear projections to the brain stem in the rat. In N. M. Gerritz, T. J. H. Ruigrok, & C. I. De Zeeuw (Eds.), *Cerebellar modules: Molecules, morphology and function* (pp. 141–172). Progress in Brain Research. Vol. 124.

Thurston, B., Buncke, J. H., Chater, L. M., & Weinstein, P. R. (1976). A scanning electron microscopy study of microarterial damage and repair. *Plastic and Reconstructive Surgery, 57*, 197–203.

Tolbert, D. L. (1978). Organizational features of the cat and monkey cerebellar nucleo-cortico projections. *Journal of Comparative Neurology, 182*, 39–56.

Tooze, J. (1964). Measurements of some cellular changes during the fixation of amphibian erythrocytes with osmium tetroxide solutions. *Journal of Cell Biology, 22*, 551–563.

Trott, J. R., Apps, R., & Armstrong, R. M. (1998). Zonal organization of cortico-nuclear and nucleo-cortical projections of the paramedian lobule of the cat cerebellum. I. The C1 zone. *Experimental Brain Research, 118*, 298–315.

Valenzuela-Chacón, J. (1970). Connections of the cerebellar granule cells with capillaries and identity of the cerebellar granule cell with neuroglia. Histochemical study and histogenesis of the superficial layer of the cerebellar granule cells. *Acta Histochemica, 37*, 302–322.

Van Gehuchten, A. (1891). La structure des centres nerveux. La moelle épinière et le cervelet. *Cellule, 7*, 83–122.

Vaquero-Crespo, J. (1975). Connections of the cerebellar grains with neuroglial and vascular structures. Interpretation of the grain as a modified neuroglial element. *Acta Anatómica, 92*, 417–423.

Voogd, J. (1967). Comparative aspects of the structure and fibre connexions of the mammalian cerebellum. *Progress in Brain Research, 25*, 94–134.

Voogd, J., & Glickstein, M. (1998). The anatomy of the cerebellum. *Trends in Neuroscience, 21*, 370–375.

Uchizono, K. (1965). Characteristic of excitatory and inhibitory synapses in the central nervous system of the cat. *Nature (London), 207*, 642–643.

Uchizono, K. (1969). Synaptic organization of the mammalian cerebellum. In: R. Llinás (Ed.), *Neurobiology of cerebellar evolution and development* (pp. 549–583). Chicago: American Medical Association/Education & Research Fund.

Weigert, C. (1895). Beiträge zur Kenntniss der normalen menschlichen Neuroglia. *Abhandlung Senckenberg Naturforschunsgesh (Frankfurt) 19*, H.II.

Westergaard, E. (1977). The blood–brain barrier to horseradish peroxidase under normal and experimental conditions. *Acta Neuropathologica (Berlin), 39*, 181–187.

Westergaard, E., & Brightman, M. W. (1973). Transport of protein across normal cerebral arterioles. *Journal of Comparative Neurology, 152*, 17–44.

Wheeler, E. E., Gavin, J. B., & Herdson, G. B. (1973). A study of endothelium using freeze drying and scanning electron microscopy. *Anatomical Record, 175*, 579–584.

Wolinsky, H. (1972). Endothelial projections. *Science, 176*, 1151.

Yamadori, T. (1972). A scanning electron microscopic observation of the choroid plexus in the rats. *Archives Histologicum Japonicum, 35*, 89–97.

Yoshida, Y., Ikuta, F., Watabe, K., & Nagata, T. (1985). Developmental microvascular architecture of the rat cerebellar cortex. *Anatomy and Embryology, 171*, 129–138.

Yu, B. P., Yu, C. C., & Robertson, R. T. (1994). Patterns of capillaries in developing cerebral and cerebellar cortices of rats. *Acta Anatómica, 149*, 128–133.

Index